CEMENT and CONCRETE
MINERAL ADMIXTURES

CEMENT and CONCRETE
MINERAL ADMIXTURES

MUSTAFA TOKYAY
Middle East Technical University,
Ankara, Turkey

CRC Press
Taylor & Francis Group
Boca Raton London New York

CRC Press is an imprint of the
Taylor & Francis Group, an **informa** business

CRC Press
Taylor & Francis Group
6000 Broken Sound Parkway NW, Suite 300
Boca Raton, FL 33487-2742

First issued in paperback 2018

ISBN: 978-1-4987-1654-3 (hbk)
ISBN: 978-0-367-02800-8 (pbk)

Library of Congress Cataloging-in-Publication Data

Names: Tokyay, Mustafa, author.
Title: Cement and concrete mineral admixtures / author, Mustafa Tokyay.
Description: Boca Raton : Taylor & Francis, 2016. | Includes bibliographical references and index.
Identifiers: LCCN 2015042113 | ISBN 9781498716543 (hard cover)
Subjects: LCSH: Concrete--Additives.
Classification: LCC TP884.A3 T65 2016 | DDC 620.1/36--dc23
LC record available at http://lccn.loc.gov/2015042113

Visit the Taylor & Francis Web site at
http://www.taylorandfrancis.com

and the CRC Press Web site at
http://www.crcpress.com

To Nuray, Ekin and Arya

Contents

Introductory remarks and acknowledgements

In the world of modern cement and concrete, mineral admixtures have become essential ingredients of these materials. The technical, economical and ecological advantages that mineral admixtures present are undisputed. The main technical reasons for their use are to enhance the workability of fresh concrete and the durability of hardened concrete. In fact, they affect almost every property of concrete. Economical and ecological reasons may be equally or sometimes more significant than technical reasons. They may enable the reduction in the amount of portland cement used, which results in both economic and ecological benefits. Furthermore, many of the mineral admixtures are industrial by-products, which may be considered as waste otherwise.

This book is intended to provide detailed information on mineral admixtures used in cement and concrete. Since there is a wide variety of mineral admixtures, some of which are in commercial use and some under development, the book is more concentrated on the commonly used ones.

The first seven chapters are related to the general characteristics of mineral admixtures, including production methods, chemical and mineralogical compositions and physical properties. Chapters 8 through 12 describe the influences of mineral admixtures on concrete properties and behaviour. The basics of each individual concrete property are also given in these five chapters so that the effects of the mineral admixtures would be better understood. Chapter 13 overviews the current international standards related to mineral admixtures in cement and concrete. Finally, in Chapters 14 and 15, the use of mineral admixtures in special concretes and cements are discussed.

There are huge volumes of reported research work related to mineral admixtures in cement and concrete. In writing this book, as many accessible papers and books as possible published upto June 2015 were taken into consideration.

The author wishes to express his gratitude to Dr. İsmail Özgür Yaman who urged (or even forced) him to write the book and made many suggestions and corrections; PhD candidate Burhan Aleessa Alam who helped him with the figures and all other computer-related work; Dr. Mustafa Şahmaran and Dr. Korhan Erdoğdu (his former PhD students) and Dr. Oğuzhan Çopuroğlu who allowed him to use some of their SEM images; and many other colleagues, friends and students with whom he shared his opinions. Special thanks are due to Turkish Cement Manufacturers' Association (TÇMB) for allowing the use of some of their research data, R&D laboratory facilities and library.

Mustafa Tokyay

Author

Dr. Mustafa Tokyay is a professor of Civil Engineering in the Middle East Technical University (METU), Ankara, Turkey. He served as the Civil Engineering Department chair (1999–2003) and dean of Engineering Faculty (2003–2006) in the same university. He worked as a visiting researcher in Dundee University, Scotland (1988–1989) and as a visiting professor in the Eastern Mediterranean University, Cyprus (1993–1994).

He has been the president of Turkish Chamber of Civil Engineers (1994–1996), member of the Executive Board of Turkish Engineering Deans' Council (2003–2006), member of the Administrative Council of SEFI (European Society of Engineering Education) (2004–2007), member of Education and Training Standing Committee of European Council of Civil Engineers (ECCE) (2006–2008).

Dr. Mustafa Tokyay was the director of the Cement and Concrete Research Department of Turkish Cement and Earthenware Industry (1995–1996), director of Cement and Concrete Research Institute of Turkish Cement Manufacturers' Association (TÇMB) (1996–1999; 2008–2011) and served in several committees and project groups of the European Cement Association (CEMBUREAU) as the representative of TÇMB (1995–2005). He has organised many national and international technical congresses and symposia on cement, concrete and mineral admixtures.

His research interests include cement and concrete technology and the use of industrial by-products in cement and concrete. He is the author, coauthor or editor of more than 70 national and international papers, proceedings and books. At the time of this book's release, Dr. Tokyay has supervised 49 MS and PhD theses.

Chapter 1

Admixtures and additions

1.1 GENERAL

An admixture is broadly defined as something that is mixed together. Many of the additional materials that are being used in cement and concrete may be considered as admixtures according to this broad definition, since they are mixed together either with cement or concrete. However, more explicit terms are being used in cement and concrete terminology. An *admixture*, as defined by the ACI Committee 116, is 'a material other than water, aggregates, hydraulic cement, and fibre reinforcement, used as an ingredient of a cementitious mixture to modify its freshly mixed, setting or hardened properties and that is added to the batch before or during mixing'. Materials of similar nature are called *additions* if they are incorporated into hydraulic cements during manufacture either by intergrinding or blending (ACI 116, 2005). Both of these terms embody a very wide range of materials some of which are commonly used while others have only limited applications.

Concrete admixtures are customarily categorised into two major groups as *chemical admixtures* and *mineral admixtures*. Chemical admixtures include air-entraining, water-reducing, high-range water-reducing, retarding and accelerating agents which are water-soluble compounds. They are added to concrete in very small amounts. The scope of chemical admixtures is much wider than that is given here. Since this book deals with commonly used mineral admixtures in cement and concrete, the reader may be referred elsewhere (Mailvaganam and Rixom, 1999; Paillére, 1995; ACI 212, 2010) for chemical admixtures.

1.2 MINERAL ADMIXTURES

Mineral admixtures are finely ground insoluble mineral matter added to concrete in large amounts. Although the terms cement replacement materials (CRM), supplementary cementitious materials (SCM) or complementary cementitious materials are sometimes synonymously used for mineral admixtures; none of them thoroughly explain the function of mineral

1

admixtures in cementitious systems. Mineral admixtures may be used to partially replace fine aggregate or as another main constituent in concrete, too. Only few of them have limited hydraulic activity of their own, in order to be called cementitious.

The main reasons for using mineral admixtures in concrete either as a direct constituent or as combined with portland cement (PC) are to enhance the workability of fresh concrete and the durability of hardened concrete. In fact, they affect almost every property of concrete and these will be discussed in the following chapters individually for the common mineral admixtures.

Besides technical reasons, economic and ecological reasons are equally or sometimes more significant in using mineral admixtures in cement and concrete. Manufacturing of PC is an energy-intensive process. About 110 kWh electric energy and about 4 GJ fuel energy is consumed to produce 1 ton of PC. On the other hand, roughly speaking, 1 ton PC production leads to 0.9 tons of CO_2 emission. Approximately half of this originates from the decomposition of the calcareous raw material for PC during the burning process. Therefore, the use of mineral admixtures in cement and concrete results in reduced energy consumption as well as lower greenhouse gas emissions since they do not require a clinkering process. Furthermore, most of these mineral admixtures are industrial by-products which when not used are wastes causing serious environmental pollution.

Mineral admixtures may be subdivided into three categories as (1) materials of low or no reactivity, (2) pozzolans and (3) latent hydraulic materials.

Those in the first category include materials like ground limestone, dolomite, quartz and hydrated lime. They are primarily used to improve the workability of fresh concrete, especially those with insufficient fine aggregate contents. Such concretes are prone to segregation and bleeding. Therefore, their coarse aggregate- and water-holding capacities (cohesiveness) should be increased by increasing the fines content.

A pozzolan is defined in the ACI Committee 216 report (ACI 116, 2005) as 'a siliceous or siliceous and aluminous material, which in itself possesses little or no cementitious value but will, in finely divided form and in the presence of moisture, chemically react with calcium hydroxide at ordinary temperatures to form compounds possessing cementitious properties'. The reaction of pozzolan with calcium hydroxide is named a pozzolanic reaction and may be written in a simple form as

$$Ca(OH)_2 + SiO_2 + H_2O \rightarrow \text{calcium silicate hydrates.} \tag{1.1}$$

Similarly, alumina that may be present in the pozzolan reacts with calcium hydroxide, under moist conditions, to form calcium aluminate hydrates:

$$Ca(OH)_2 + Al_2O_3 + H_2O \rightarrow \text{calcium aluminate hydrates.} \tag{1.2}$$

Reactions 1.1 and 1.2 take place if the pozzolan is non-crystalline (amorphous or glassy). A pozzolan may be directly mixed with calcium hydroxide but when it is used with PC, it reacts with the calcium hydroxide formed upon the hydration of PC. Pozzolanic reactions are almost always gradual.

A wide variety of materials that contain silica or silica and alumina have pozzolanic properties. They may be naturally occurring materials such as volcanic ashes, tuffs, diatomaceous earth and pumicite or industrial by-products such as low-lime fly ash (LLFA), silica fume (SF) and rice husk ash (RHA). Crushing, grinding and sometimes size separation are the processes generally used for natural pozzolans. Clays and shales need to be thermally activated at 600–900°C and then rapidly cooled in order to attain pozzolanic properties. Thus, their microstructure changes from crystalline to amorphous (Massazza, 1988). By-product materials do not usually require processing, however, drying and further particle size reduction may sometimes be needed.

Common uses of pozzolans in concrete comprise workability improvement, lowering the heat of hydration and heat evolution rate, and improving many durability properties. Pozzolans also contribute to later age strength of concrete.

Latent hydraulic materials have the ability to form cementitious products upon reacting with water. However, the reaction is much slower than those of PCs and usually requires a chemical activator of alkaline or sulphatic origin (Odler, 2000). This group of mineral admixtures consist of

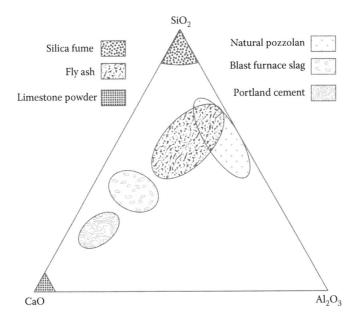

Figure 1.1 Ternary diagram of common mineral admixtures.

granulated blast furnace slags (GBFS) and high-lime fly ashes (HLFA). The former is a by-product of pig iron production. It leaves the blast furnace in molten form. In order to be used as a mineral admixture in concrete or addition in cement, it should be cooled rapidly so that it will be composed mostly of a glassy phase. The crystalline content of a high quality GBFS is lower than 5%. Its glass phase is mainly silicate with calcium, aluminium and magnesium cations (Moranville-Regourd, 1988). GBFS has to be ground to at least cement fineness. High-lime fly ashes are obtained as a by-product of thermal power plants burning lignitic and sub-bituminous coals. They contain 30%–80% silicate glass with calcium, magnesium, aluminium, sodium and potassium ions (Alonso and Wesche, 1991; Mehta and Monteiro, 2006). The crystalline matter contains quartz, free lime and free magnesium oxide. Anhydrite may be present in HLFA if high sulphur coal is used in the power plant (Mehta and Monteiro, 2006). GBFS and HLFA are both latent hydraulic and pozzolanic.

$CaO–Al_2O_3–SiO_2$ ternary diagram of common mineral admixtures, in comparison with PC is given in Figure 1.1.

Chapter 2

Natural pozzolans

2.1 GENERAL

The term pozzolan originates from the ancient Roman town of Puteoli (Pozzuoli), the glassy pyroclastic rocks of which had long been known to react with lime, in powdered form, under moist conditions to attain binding characteristics. Today, however, pozzolan has a much wider definition that embodies all those natural and artificial inorganic materials that, in pulverised state and in the presence of water, react with calcium hydroxide to form compounds of cementitious value. This broad definition embraces many different materials in terms of origin, composition and structure. A precise classification of pozzolans is almost impossible, as the broad definition includes so many different materials. However, pozzolanic materials can be divided into two major groups as natural pozzolans and artificial pozzolans. The basic difference between the two groups is that the former do not need any treatment except grinding whereas the latter are either obtained by the heat treatment of materials with no or very little pozzolanicity or as industrial by-products which experienced high temperatures and then sudden cooling during the industrial processes from which they are obtained.

A fairly comprehensive classification of pozzolans which was proposed by F. Massazza (1974) and updated with small modifications later (Massazza, 1988) is given in Figure 2.1.

Various artificial pozzolans will be discussed later in the relevant chapters. This chapter will be devoted to natural pozzolans.

Natural pozzolans can be subdivided into two categories as materials of volcanic origin (pyroclastic) and materials of sedimentary origin (clastic). The first category includes materials formed by the quenching of molten magma when it is projected to the atmosphere upon explosive volcanic eruptions. The explosive eruption has two consequences: (1) The gases, originally dissolved in the magma are released by the sudden decrease of pressure. This causes a microporous structure in the resulting material. (2) Rapid cooling of the molten magma particles when contacted with the atmosphere results in quenching, which is responsible for the glassy state of the solidified material. Pozzolanic materials of volcanic origin may be

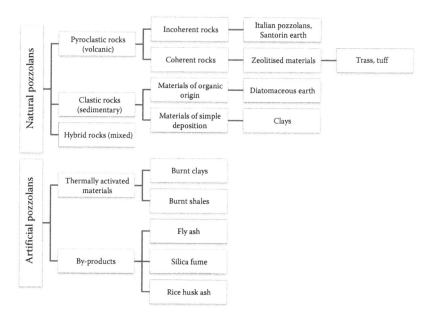

Figure 2.1 Classification of pozzolans. (From *Lea's Chemistry of Cement and Concrete*, 4th Ed., Massazza, F., Pozzolana and pozzolanic cements, Copyright 1988, with permission from Elsevier.)

found in loose (incoherent) or compacted (coherent) forms in nature. The latter results from the post-depositional processes such as weathering, compaction, cementation and hardening of the originally loose material. These processes may change the original structure into clayey or zeolitic character. Transformation into clayey structure reduces the pozzolanicity whereas zeolitisation improves it (Massazza, 1988).

The second category of natural pozzolans includes clays and diatomaceous earth. Clays have very limited pozzolanic reactivity unless they are thermally treated. Diatomaceous earth, which is a sedimentary rock, consists basically of the fossilised remains of diatoms (a type of algae). It has an amorphous siliceous structure but may contain crystalline phases upto 30%, by mass (Aruntaş et al., 1998).

Scanning electron microscopy images of several natural pozzolans are given in Figure 2.2.

2.2 CHEMICAL COMPOSITION

The chemical compositions of natural pozzolans vary widely with silica (SiO_2) prevailing over the other constituents. Alumina (Al_2O_3) has the second highest mass and then comes iron oxide (Fe_2O_3). In most of the standards related to the use of natural pozzolans in cement and concrete, the

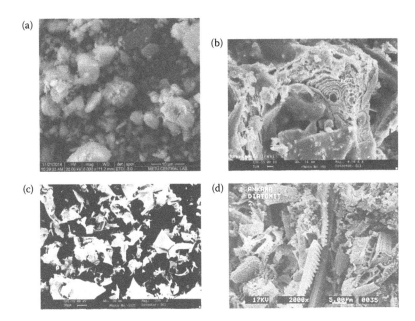

Figure 2.2 SEM images of various natural pozzolans: (a) volcanic ash. (Courtesy of Ilker Tekin.); (b) trass. (Courtesy of TÇMB R&D Institute, Ankara, Turkey.); (c) raw perlite. (Courtesy of TÇMB R&D Institute, Ankara, Turkey.) and (d) diatomite. (Reprinted with permission from Aruntaş, H.Y. 1996. Usability of diatomites as pozzolans in cementitious systems, PhD thesis, Gazi University Institute of Natural and Applied Sciences [In Turkish].)

sum of these three oxides is limited to a minimum (70%, by mass) in order for the pozzolan to be a suitable one, besides several other requirements which will be discussed in detail in Chapter 13.

Examples for the chemical compositions of various incoherent and coherent pyroclastic pozzolanic materials and diatomaceous earths are given in Table 2.1. The chemical compositions of pyroclastic materials depend on the composition of the ejected magma. Diatomaceous earth is very rich in silica since it is formed from the siliceous skeletons of diatoms.

2.3 MINERALOGICAL COMPOSITION

Similar to the variability in chemical compositions, many different minerals may be present in natural pozzolans. However, the fundamental constituent is always a glassy phase of siliceous nature which can be distinguished by the halo corresponding to about 23° 2θ (Cu Kα) that is close to the main peaks of various crystalline forms of silica in the x-ray diffractograms. Glass content of natural pozzolans of volcanic origin generally varies from 50% to 97% and the rest are mostly clay minerals, quartz

Table 2.1 Chemical compositions of various natural pozzolans (%)

Type	Reference	SiO_2	Al_2O_3	Fe_2O_3	CaO	MgO	Na_2O	K_2O	TiO_2	SO_3	LOI
Pyroclastic (incoherent)	Costa and Massazza (1974)	53.08	17.89	4.29	9.05	1.23	3.08	7.61	0.31	0.65	3.05
	Takemoto and Uchikawa (1980)	71.77	11.46	1.14	1.10	0.54	1.53	2.55	0.14	–	6.50
	Meral (2004)	76.57	9.99	0.96	0.51	0.03	–	5.58	–	0.04	5.23
	Erdoğdu (1996)	48.52	17.49	7.80	7.84	1.41	5.20	3.10	2.02	0.29	1.75
Pyroclastic (coherent)	Ludwig and Schwiete (1963)	52.12	18.29	5.81	4.94	1.20	1.48	5.06			11.10
	Kasai et al. (1992)	71.65	11.77	0.81	0.88	0.52	1.80	3.44		0.34	9.04
	Erdoğdu (1996)	57.74	12.14	3.84	4.84	0.55	0.64	2.50	0.73	0.30	11.36
	Massazza (1988)	54.68	17.70	3.82	3.66	0.95	3.43	6.38			9.11
Clastic (diatomaceous earth)	Johansson and Andersen (1990)	75.6	8.62	6.72	1.10	1.34	0.43	1.42		1.38	2.15
	Mielenz et al. (1950)	85.97	2.30	1.84		0.61	0.21	0.21			8.29
	Tonak (1995)	84.24	4.75	0.91	0.94	0.26	0.15	0.25		0.09	8.47
	Aruntaş et al. (1998)	88.32	3.47	0.48	0.42	0.26	0.17	0.28	0.18		5.84

Table 2.2 Minerals in various natural pozzolans

Type	Reference	Mineral phases
Pyroclastic (incoherent)	Costa and Massazza (1974)	Glass, feldspars, quartz, olivine, clay minerals
	Mehta (1981)	Glass, quartz, anorthite, labradorite
	Erdoğdu (1996)	Glass, augite, albite, hornblende
	Mielenz et al. (1950)	Glass, calcite, quartz, feldspar, sanidine, montmorillonite
Pyroclastic (coherent)	Ludwig and Schwiete (1963)	Glass, quartz, feldspar
	Erdoğdu et al. (1999)	Glass, quartz, calcite, albite, sanidine
	Erdoğdu (1996)	Glass, clinoptilolite, quartz, sanidine, illite
Clastic (diatomaceous earth)	Aruntaş et al. (1998)	Glass, quartz, feldspar, smectite
	Takemoto and Uchikawa (1980)	Opal, quartz, cristobalite

and feldspars. Glassy phase in diatomaceous earths may be as low as 25% and some may be almost totally glassy. The remainder is composed of clay minerals, quartz and feldspars. Minerals in some natural pozzolans are listed in Table 2.2 and examples of the x-ray diffractograms are shown in Figure 2.3.

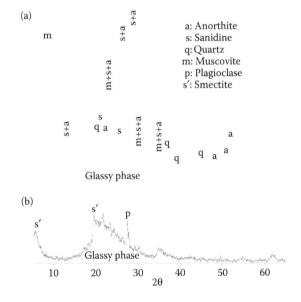

Figure 2.3 X-ray diffractograms of two natural pozzolans: (a) trass. (Courtesy of TÇMB R&D Institute, Ankara, Turkey.) and (b) diatomaceous earth. (With permission from Aruntaş, H.Y. 1996. Usability of diatomites as pozzolans in cementitious systems, PhD thesis, Gazi University Institute of Natural and Applied Sciences [In Turkish].)

2.4 FINENESS

Natural pozzolans are used in cementitious systems after they are finely ground to a desired fineness which generally lies between 350 and 600 m^2/kg (Blaine specific surface). When they are used as a component in blended cements, they are usually interground with PC clinker.

Chapter 3

Fly ashes

3.1 GENERAL

Coal is a major fuel for energy production in coal-burning thermal power plants throughout the world. In the first decade of the twenty-first century, around 40% of global electricity was supplied by coal burning (Heidrich et al., 2013). Several different coal combustion products (CCP) such as fly ash (FA), bottom ash, boiler slag and flue gas desulphurisation gypsum are obtained by burning pulverised coal in thermal power plants. According to 2010 figures, the total annual CCP production had reached about 800 Mt (Heidrich et al., 2013). Around 70% of this total CCP obtained is FA (Von Berg and Feuerborn, 2005).

FA is one of the by-products of coal-burning thermal power plants. Pulverised coal is used to heat water in order to drive the steam generators that convert thermal energy into electricity (Alonso and Wesche, 1991). For this purpose different types of coals such as lignite, sub-bituminous and bituminous coals and anthracite may be used. Besides carbon and some volatile matter, coals contain impurities such as clay, shale, quartz, feldspar, etc. During the burning process the ash formed from the impurities remain suspended in the flue gases. Some of them coagulate and precipitate to form the *bottom ash* whereas the rest is transported by the gases and collected by mechanical separators or electrostatic precipitators. The latter is called *fly ash* (Tokyay, 1987). Combustion system of a typical coal-burning thermal power plant is schematically shown in Figure 3.1.

The total amount of ash resulting from coal combustion in a thermal power plant may change from 6% to 40% (by mass) of the coal used. Anthracites have 6%–15% ash whereas lignites have 20%–40%. From a typical 1,000 MW thermal power plant, approximately 650,000 t of FA and bottom ash is obtained, annually (Tokyay and Erdoğdu, 1998).

FA is a fine-grained solid material having a particle size range of 0.2–200 µm. The range of particle size of any given FA depends largely on the fineness of the pulverised coal and the type of collection equipment used in the thermal power plant. FA collected by mechanical separators is coarser than that collected by electrostatic precipitators. The colour of FA may vary

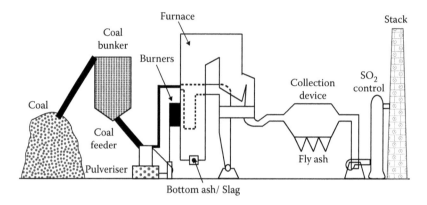

Figure 3.1 Simplified combustion system of a typical coal-burning thermal power plant. (Adapted from Whitfield, C.J. 2003. The production, disposal, and beneficial use of coal combustion products in Florida, MS thesis, University of Florida.)

from light tan to brown and from grey to black, depending on the minerals present in it. In bulk, it is generally greyish and may become darker with an increasing amount of unburnt carbon.

The chemical composition of FA is determined by the types and relative proportions of the mineral matter present in the coal used. More than 85% of most FA consists of silica (SiO_2), alumina (Al_2O_3), iron oxide (Fe_2O_3), lime (CaO) and magnesia (MgO). Generally, FA from the combustion of lignitic and sub-bituminous coals contains more lime than FA from bituminous coals and anthracite.

Owing to their fine particle sizes and generally non-crystalline character, fly ashes usually show satisfactory pozzolanic property or, in the case of having high lime content, both pozzolanic and latent hydraulic properties. The recognition of their pozzolanic character as early as 1910s (Anon, 1914) has led to their use as an important ingredient in concrete since 1930s.

The chemical and mineralogical compositions as well as the physical properties of fly ashes vary depending on the source of the coal used in the thermal power plant, method of burning, combustion equipment, collection methods, etc. Therefore, one can expect relatively large variations between fly ashes obtained from different thermal power plants and even within the same power plant at different times.

3.2 MINERALOGICAL COMPOSITION

Rapid cooling of FA, when it leaves the furnace of the power plant, results in 50%–85% of it becoming amorphous. Although there are three different types of burning processes such as (i) high-temperature combustion, (ii) dry combustion and (iii) fluidised-bed combustion employed in coal-burning thermal power plants, the second process is more common (Alonso

Table 3.1 Minerals in fly ashes

Mineral		Amount range (%)	
Name	Formula	Fly ashes from anthracites	Fly ashes from lignites
Quartz	SiO_2	2.0–8.5	2.0–6.0
Mullite	$Al_6Si_2O_{13}$	6.5–14.0	1.0–5.0
Hematite	Fe_2O_3	1.5–7.0	2.0–6.0
Magnetite	Fe_3O_4	0.5–10.0	0.6–2.5
Lime	CaO	≤3.5	5.0–30.0
Anhydrite	$CaSO_4$	Nil	0.7–16.9
Plagioclase[a]	$NaAlSi_3O_8$–$CaAl_2Si_2O_8$	Nil	Nil–28.0
Amorphous	–	45.0–95.4	30.0–80.0

Sources: Adapted from Alonso, J.L., Wesche, K. 1991. Characterization of fly ash, in *Fly Ash in Concrete Properties and Performance* (Ed. K. Wesche), E.&F.N. Spon, London; Hubbard, F.H., Dhir, R.K. 1984. *Materials Science Letters*, 3, 958–960; Tokyay, M., Hubbard, F.H. 1992. Mineralogical investigations of high-lime fly ashes, *4th CANMET/ACI International. Conference on. FlyAsh, Silica Fume, Slag and Natural Pozzolans in Concrete* (Ed. V.M. Malhotra), Vol. 1, pp. 65–78, Istanbul.

[a] A solid solution with a composition ranging from albite ($NaAlSi_3O_8$) to anorthite ($CaAl_2Si_2O_8$).

and Wesche, 1991). The burning temperature in the dry combustion process is between 1,000°C and 1,400°C. The mineral phases of the fly ashes obtained at different burning temperatures even within this range may show considerable variations both in type and amount (Hubbard and Dhir, 1984; Tokyay and Erdoğdu, 1998). On the other hand, the impurities present in the coal are reflected in the mineralogical composition of FA. Common mineral phases present in the fly ashes obtained from anthracites and lignites are given with their amount ranges in Table 3.1.

3.3 CHEMICAL COMPOSITION AND CLASSIFICATION

Even though the constituents in fly ashes are seldom in free oxide form, oxide analysis is traditionally used to describe their chemical compositions. The chemical compositions of fly ashes may show large variations, like their mineralogical compositions, depending on the type and amount of impurities present in the coal, combustion method and collection method. Fly ashes obtained from anthracite are rich in silica (SiO_2), alumina (Al_2O_3) and iron oxide (Fe_2O_3), whereas fly ashes of lignitic origin may contain significant amounts of lime (CaO) and sulphur trioxide (SO_3) in addition to the formers. Bituminous and sub-bituminous coals generally result in fly ashes in between. Other oxides such as MgO, MnO, TiO_2, Na_2O, K_2O, etc. may also appear in fly ashes in smaller amounts. The range of variations in the chemical compositions of fly ashes obtained from different coals is given in Table 3.2.

Table 3.2 Chemical composition ranges of fly ashes obtained from different types of coals

Oxide (%)	Coal type from which the fly ashes are obtained			
	Anthracite	Bituminous	Sub-bituminous	Lignite
SiO_2	47–68	7–68	17–58	6–45
Al_2O_3	25–43	4–39	4–35	6–23
Fe_2O_3	2–10	2–44	3–19	1–18
CaO	0–4	1–36	2–45	15–44
MgO	0–1	0–4	0–8	3–12
Na_2O	–	0–3	–	0–11
K_2O	–	0–4	–	0–2
SO_3	0–1	0–32	3–16	6–30

Source: Adapted from Alonso, J.L., Wesche, K. 1991. Characterization of fly ash, in Fly Ash in Concrete Properties and Performance (Ed. K. Wesche), E.&F.N. Spon, London.

There have been a number of attempts to classify fly ashes, some of which are given here. All these classifications are basically made according to the chemical composition.

3.3.1 Classifications according to SiO_2, Al_2O_3, CaO and SO_3 contents

One of the early classifications has three groups of fly ashes such as (a) silicoaluminous, (b) sulphocalcic and (c) silicocalcic. Silicoaluminous fly ashes are composed mostly of SiO_2 and Al_2O_3 and they are obtained from the combustion of anthracite and some bituminous coals. Fly ashes rich in SO_3 and CaO are called sulphocalcic. Such fly ashes are generally obtained from lignites of lower calorific value. Silicocalcic fly ashes contain high amounts of SiO_2 and CaO and they are obtained from sub-bituminous coals and lignites. Chemical compositions and x-ray diffractograms of the three fly ashes representing each of these classes are given in Table 3.3 and Figure 3.2, respectively.

Table 3.3 Chemical compositions of typical silicoaluminous, sulphocalcic and silicocalcic fly ashes

Oxide (%)	Silicoaluminous FA	Sulphocalcic FA	Silicocalcic FA
SiO_2	59.4	17.1	46.1
Al_2O_3	25.8	9.0	26.9
Fe_2O_3	5.8	3.9	4.4
CaO	2.0	49.2	15.3
MgO	0.6	1.7	0.6
Na_2O	0.4	0.2	0.3
K_2O	3.8	0.4	2.0
SO_3	0.1	14.8	2.4

Source: Adapted from Tokyay, M., Erdoğdu, K. 1998. Characterization of Turkish Fly Ashes, TÇMB AR-GE/Y 98.3,, Turkish Cement Manufacturers' Association, 70pp. [in Turkish].

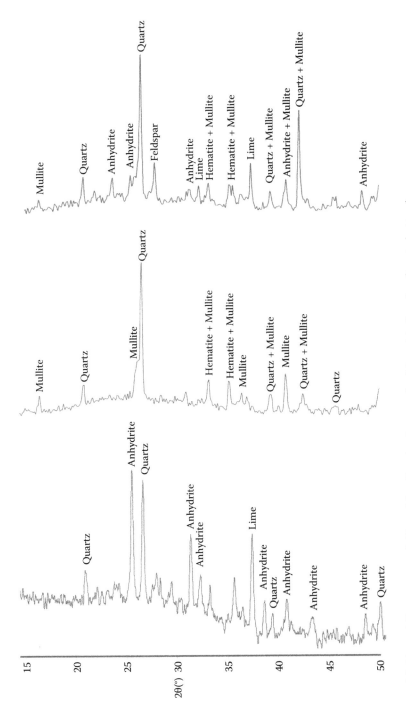

Figure 3.2 X-ray diffractograms of silicocalcic, silicoaluminous and sulphocalcic fly ashes (top to bottom).

Table 3.4 Classification of fly ashes according to their main constituents

Constituent	Amount (%)			
	Type I	Type II	Type III	Type IV
SiO_2	>50	35–50	<35	Very low
Al_2O_3		High		
Fe_2O_3		Medium		
$Al_2O_3 + Fe_2O_3$	Medium			
CaO	<7	More than Type I	Very high	Very high

Source: Adapted from Alonso, J.L., Wesche, K. 1991. Characterization of fly ash, in *Fly Ash in Concrete Properties and Performance* (Ed. K. Wesche), E.&F.N. Spon, London.

A similar classification which groups fly ashes into four as Type I, II, III and IV according to the percentage of main constituents is summarised in Table 3.4 (Alonso and Wesche, 1991).

3.3.2 Classifications according to CaO content

A broad classification, with particular regard to their use in cement and concrete, divides fly ashes into two categories as low-lime and high-lime fly ashes (Dhir, 1986). The dividing line between these two categories is 10% CaO content. Low-lime fly ashes are pozzolanic whereas high-lime fly ashes possess some cementitious value in addition to pozzolanic property.

This classification is found to be reflected to a certain extent in two important international standards: ASTM specification for FA (ASTM C 618, 2012) defines two groups of fly ashes as Class F and Class C. They are distinguished on the basis of the coal type from which the FA is obtained. Class F fly ashes are generally obtained from anthracite or bituminous coals and Class C fly ashes are generally obtained from sub-bituminous coals or lignites. Besides the type of coal, it is required by the standard that the sum of SiO_2, Al_2O_3 and Fe_2O_3 contents should be ≥70% in Class F, and ≥50% in Class C fly ashes. Furthermore, ASTM C 618 states that some Class C fly ashes may contain more than 10% CaO.

European cement standard, EN 197-1 (2012) considers fly ashes as a major constituent (besides PC clinker) of several common cements. The standard groups fly ashes into two as siliceous (V) and calcareous (W). V-type fly ashes should contain at least 25% (by mass) reactive SiO_2 and at most 10% reactive CaO. W-type fly ashes, on the other hand, are the ones that contain more than 10% reactive CaO and the reactive silica content should not be less than 25%. By definition, reactive silica is the silica of the FA which is treated with hydrochloric acid that is soluble in boiling potassium hydroxide solution (KOH) and reactive lime is total lime content minus the lime of $CaCO_3$ and $CaSO_4$ that is present in the FA.

Some of the international standards on the use of fly ashes in cement and concrete will be discussed in more detail in Chapter 13.

3.4 FINENESS, PARTICLE SIZE DISTRIBUTION AND PARTICLE MORPHOLOGY

FA fineness is generally determined by sieve analysis although other techniques such as Blaine air permeability, hydrometer analysis, nitrogen adsorption and laser particle size diffraction may also be employed. Most national and international standards specify the maximum amount of material retained on a 45 μm sieve upon wet sieving. This method is simple, fairly reliable and cheap. ASTM C 618 limits this amount to maximum 34% (by mass), whereas the European standard on FA for concrete, EN 450-1(2008) specifies two categories of fly ashes as N and S with maximum 40% and 12% retained on a 45 μm sieve, respectively. FA fineness in terms of specific surface area lies between 250 and 550 m^2/kg.

Although it is generally accepted that FA consists mostly of spherical particles, FA particle morphology depends largely on the type and fineness of the coal from which it is obtained and the combustion conditions and collection method used in the thermal power plant. Consequently, they may be spherical, angular or irregular in shape with particle surfaces being smooth, rough or coated. Lower combustion temperatures (≤1,100°C) generally lead to irregular particles since the mineral impurities in the coal fail to melt (Alonso and Wesche, 1991). These particles consist of minerals such as quartz, feldspars, spent clay and unburnt carbon. Most of the impurities in the coal melt at high combustion temperatures (>1,500°C) and result in spherical particles. These may be solid glassy spheres, hollow spheres (cenospheres), spheres containing a number of smaller spheres (plerospheres), or spheres with crystalline matter covered surfaces (dermaspheres) (Alonso and Wesche, 1991; Türker et al., 2004). Quartz particles which are able to withstand the high temperatures and newly formed crystalline matter like mullite may be observed in these fly ashes, also.

Low-lime fly ashes, when compared with the high-lime ones have a more homogeneous particle structure composing mostly of spherical particles. High-lime fly ashes contain a considerable amount of angular and irregular particles of minerals such as anhydrite, free lime, calcite and feldspars besides the spherical ones. The examples of various FA particle shapes are given in Figure 3.3.

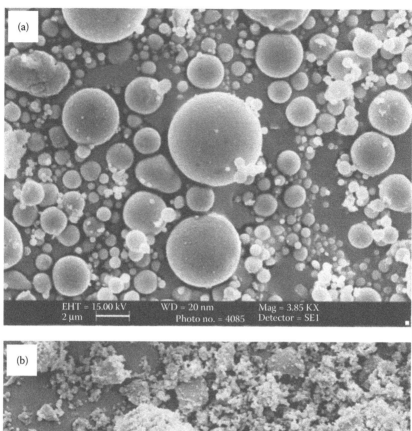

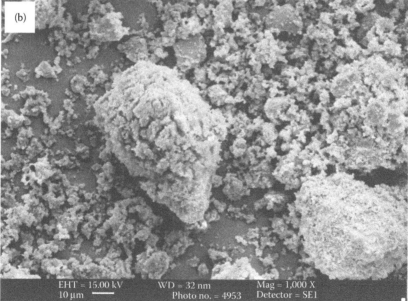

Figure 3.3 SEM images of (a) cenospheres and solid spheres and (b) massive free lime. (Adapted from Türker, P. et al. 2004. Classification and Properties of Turkish Fly Ashes, TÇMB/AR-GE/Y03.03, Turkish Cement Manufacturers' Association, 99pp. [in Turkish]. With permission.)   *(Continued)*

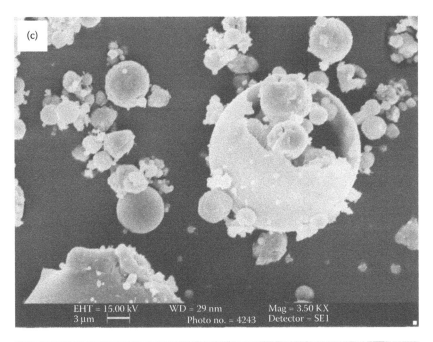

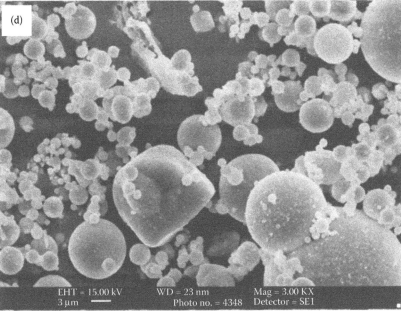

Figure 3.3 (Continued) SEM images of (c) plerosphere and (d) angular quartz crystal. (Adapted from Türker, P. et al. 2004. Classification and Properties of Turkish Fly Ashes, TÇMB/AR-GE/Y03.03, Turkish Cement Manufacturers' Association, 99pp. [in Turkish]. With permission.)   (Continued)

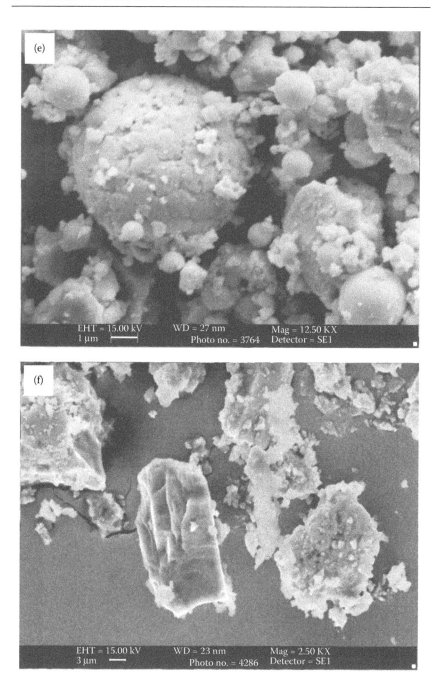

Figure 3.3 *(Continued)* SEM images of (e) magnetite covered sphere and (f) calcite crystal. (Adapted from Türker, P. et al. 2004. Classification and Properties of Turkish Fly Ashes, TÇMB/AR-GE/Y03.03, Turkish Cement Manufacturers' Association, 99pp. [in Turkish]. With permission.)   *(Continued)*

Figure 3.3 (Continued) SEM images of (g) unburnt carbon with microspheres and (h) agglomerated mullite crystals. (Adapted from Türker, P. et al. 2004. Classification and Properties of Turkish Fly Ashes, TÇMB/AR-GE/Y03.03, Turkish Cement Manufacturers' Association, 99pp. [in Turkish]. With permission.)

3.5 DENSITY

The average relative density of FA ranges between 1,900 and 2,400 kg/m³ (Dhir, 1986; Alonso and Wesche, 1991). Density may be as high as 2,900 kg/m³ when the FA is ground until no hollow particle remains (Dhir, 1986). Different particle size ranges may have different densities ranging from 500 to 2,600 kg/m³. Higher densities were found to be correlated with the increasing amount of iron oxide in the FA. Loose bulk unit weight of dry FA ranges between 500 and 900 kg/m³.

3.6 RADIOACTIVITY

Minerals, rocks, coals, etc. may possess some radioactivity. It results from the presence of very small quantities of radioactive elements such as radium (^{226}Ra), thorium (^{232}Th) and an isotope of potassium (^{40}K) whose nuclei disintegrate to give off corpuscular or electromagnetic radiation. Radioactivity is commonly determined as the number of disintegrations per unit mass of the unstable element per unit time. The SI unit for radioactivity is Becquerel (Bq = disintegration/s).

Radium equivalent activity index (Ra_{eq}) is a common measure of the radioactivity. It is based on the fact that 1 kg of ^{226}Ra, ^{232}Th and ^{40}K result in 10, 7 and 130 Bq, which produce the same gamma dose rate, respectively. It is calculated as

$$Ra_{eq} = C_{Ra} + \frac{10}{7}C_{Th} + \frac{10}{130}C_{K} \qquad (3.1)$$

where C_{Ra}, C_{Th} and C_{K} are the specific activities of these radioactive elements in Bq/kg (Turhan et al., 2010). The generally accepted Ra_{eq} for building materials is 370 Bq/kg (NEA-OECD, 1979). Higher values such as 2,200 Bq/kg for road constructions and 3,700 Bq/kg for non-residential land filling may be acceptable (Turhan et al., 2010).

A recent study revealed that the mean Ra_{eq} values range from 241.5 to 1,102.1 Bq/kg for 15 Turkish fly ashes (Turhan et al., 2010). In a previous study, the radioactivity of mortars in which the cements were partially replaced by fly ashes obtained from nine different thermal power plants in Turkey, Ra_{eq} values were found to be in the range of 26.6–87.6 Bq/kg, 40.5–115.6 Bq/kg and 49.2–172.2 Bq/kg, respectively for 10%, 20% and 30% (by mass) FA incorporations (Tokyay and Erdoğdu, 1998). Therefore, it can be stated that radioactivity of the actual construction material in which the FA is used may need to be determined in addition to that of the FA.

Chapter 4

Blast furnace slag

4.1 GENERAL

In cement and concrete jargon the term slag usually means blast furnace slag (BFS) which is a by-product of pig iron production. Actually, in the general sense of the term, slags are by-products of pyro-metallurgical processes in the metal and alloy industries. Their compositions and amounts depend on the process by which the slag is formed and on the materials used in the process. Typical chemical compositions of various metallurgical slags are given in Table 4.1.

Although there are numerous research works on the use of many different metallurgical slags in cement and concrete as cementitious or pozzolanic materials or aggregate, their use is very little except for iron BFS. BFS is the most commonly used one in the cement and concrete industry since the 1800s (Lewis et al., 2003).

Production of pig iron is carried out in blast furnaces into which controlled amounts of iron ore, coke (as fuel) and limestone (as flux) are fed. The mixture is subjected to intense heat by means of high-pressure hot air blasts through a series of jets located near the bottom of the furnace. The main reactions in the blast furnace can be shown in a simplified manner as follows:

1. The hot air blown into the furnace reacts with the carbon in the form of coke to produce carbon monoxide and heat:

$$C + O_2 \rightarrow CO_2 \tag{4.1}$$

$$CO_2 + C \rightarrow 2CO \tag{4.2}$$

2. Carbon monoxide then reacts with iron oxide of the ore to produce iron:

$$Fe_2O_3 + 3CO \rightarrow 2Fe + 3CO_2 \tag{4.3}$$

Table 4.1 Typical chemical compositions of various metallurgical slags

Oxide (%)	Ferrous and ferroalloy industry slags					Non-ferrous metal industry slags			
	Iron blast furnace slag (Tokyay and Erdoğdu, 1998)	Steel basic oxygen furnace slag (Zhang et al., 2012)	Steel electric arc furnace slag (Iacobescu et al., 2011)	Ferronickel slag (Dourdounis et al., 2004)	Ferrochromium slag (Zelić, 2005)	Copper slag (Kıyak et al., 1999)	Phosphorus slag (Xu et al., 2012)	Zinc slag (Weeks et al., 2008)	Lead slag (Saikia et al., 2008)
SiO_2	41.6	14.8	18.1	34.3	33.4	31.9	39.0	18.9	25.7
CaO	32.7	39.1	32.5	14.1	2.1	4.0	47.5	13.9	18.9
MgO	13.9	9.3	2.5	2.0	33.23	2.8	3.6	2.1	1.4
Al_2O_3	1.1	3.3	13.3	5.7	15.4	2.4	2.5	8.5	5.4
Fe_2O_3	7.0	25.1	26.3	37.6	2.6	39.7	1.0	39.2	34.0
SO_3	1.6	0.2	0.4	–	0.5	–	0.1	3.8	–
Na_2O	0.8	–	0.1	–	–	–	0.1	1.0	0.9
K_2O	0.9	–	–	–	–	–	0.2	0.6	0.3
P_2O_5	–	–	–	–	–	–	3.7	0.4	–
TiO_2	0.1	–	0.5	–	–	–	–	0.4	–
Cr_2O_3	–	–	–	3.5	13.5	–	–	0.2	–

3. While obtaining the pig iron, the limestone decomposes into lime and carbon dioxide in the middle zones of the furnace:

$$CaCO_3 \rightarrow CaO + CO_2 \qquad (4.4)$$

4. The lime formed reacts with the impurities (notably the siliceous impurities) in the iron ore resulting in a molten slag which is essentially calcium silicate:

$$CaO + SiO_2 \rightarrow CaSiO_3 \qquad (4.5)$$

A simplified schematic representation of pig iron production is shown in Figure 4.1.

Thus, another important outcome of a blast furnace, besides the pig iron, is the slag. Typically, 220–370 kg of BFS is obtained per tonne of pig iron production. Sometimes, lower grade ores yield much higher amounts of slag (Kalyoncu, 2000). The slag is in molten form at a temperature between 1,400°C and 1,600°C when it comes out of the blast furnace. Most of its characteristics are affected by its chemical composition and rate of cooling. Depending on the cooling method, three types of BFSs are obtained: air-cooled, expanded and granulated. Air-cooled slags solidify into a dense crystalline structure composed of Ca–Al–Mg-silicates. They can be used as concrete aggregate.

Expanded slags are formed by rapid cooling in water with steam and compressed air application. Steam and air applications result in a cellular, porous structure. Granulated slags are formed by applying pressurised air or water that turns the molten slag into small droplets which are then quenched in

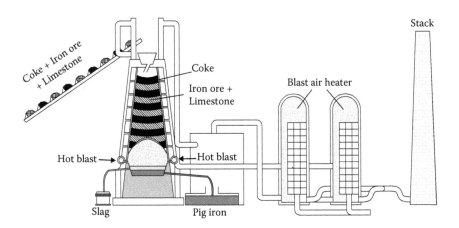

Figure 4.1 Simplified schematic representation of iron blast furnace.

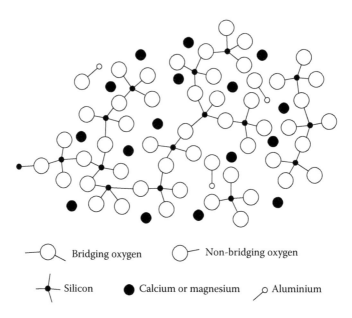

Bridging oxygen

Non-bridging oxygen

Silicon

Calcium or magnesium

Aluminium

Figure 4.2 Schematic internal structure of GBFS. (Adapted from Regourd, M. 1986. Slags and slag cements, in *Cement Replacement Materials* [Ed. R.N. Swamy], Surrey University Press.)

water. Expanded and granulated slags can also be obtained by first applying high-pressure water jets and then flinging them into the air in a rotary drum. Expanded slags of size 4–15 mm and granulated slags of size less than 4 mm are obtained separately. Expanded slags have highly porous and partly crystalline structure that makes them suitable as an aggregate in lightweight concrete. Granulated slags are almost completely glassy and attain latent hydraulic and pozzolanic properties upon pulverising (Moranville-Regourd, 1988; Kalyoncu, 2000; Lewis et al., 2003). Ground granulated slags are suitable as a mineral admixture in cement and concrete.

The internal structure of GBFS can be explained in a convenient manner by considering it as a supercooled liquid silicate in which some of the Si–O–Si bonds are broken and Al^{3+} ions replace some of Si^{4+} ions (Moranville-Regourd, 1988). In order to restore the overall electrical neutrality Ca^{2+} and Mg^{2+} cations are placed in the cavities of the network (Regourd, 1986; Moranville-Regourd, 1988; Tokyay and Erdoğdu, 1998). The resulting structure is shown in Figure 4.2.

4.2 MINERALOGICAL COMPOSITION

GBFS is almost completely glassy. However, small amounts of crystal phases such as merwinite $(Ca_3Mg(SiO_4)_2)$, melilite (a calcium–aluminium–magnesium

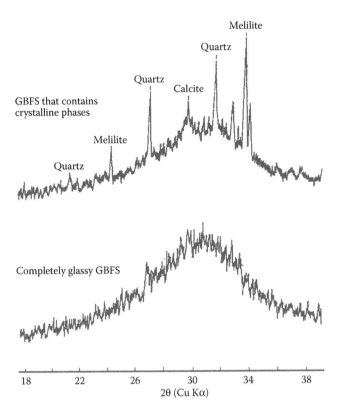

Figure 4.3 Typical x-ray diffractograms of GBFS. (Courtesy of TÇMB.)

silicate solid solution with a composition ranging from gehlenite ($Ca_2Al(AlSi)O_7$) to akermanite ($Ca_2MgSi_2O_7$), calcite and quartz. The glassy character of GBFS is manifested by a broad hunch with a peak corresponding to approximately 30° 2θ in Cu Kα x-ray diffractograms as shown in Figure 4.3.

4.3 CHEMICAL COMPOSITION

From the view point of its suitability for use in cement and concrete, the most important parameter is the glass content of the GBFS. Increasing amount of crystalline components in it would reduce its pozzolanic and cementitious value. The chemical composition may also be important in evaluating the hydraulic reactivity of a GBFS. Generally, the more basic the slag, the higher will be its reactivity. It was observed that the reactivity of GBFS increases with increasing CaO, Al_2O_3 and MgO contents and decreases with increasing SiO_2 content (Pal et al., 2003). There are several basicity indices or hydraulic moduli proposed by different researchers to assess the hydraulic reactivity of GBFS as given in Table 4.2. These indices were determined

Table 4.2 Basicity indices or hydraulic moduli proposed for assessing the hydraulicity of GBFS

No	Index or modulus	Requirement for good performance
1	$\dfrac{CaO}{SiO_2}$	0.3–1.4
2	$\dfrac{CaO + MgO}{SiO_2}$	>1.4
3	$\dfrac{CaO + MgO + \frac{2}{3}Al_2O_3}{SiO_2 + \frac{1}{3}Al_2O_3}$	>1.0
4	$\dfrac{CaO + MgO + Al_2O_3}{SiO_2}$	>1.0
5	$\dfrac{CaO + MgO}{SiO_2 + Al_2O_3}$	1.0–1.3
6	$\dfrac{CaO + 0.56Al_2O_3 + 1.40MgO}{SiO_2}$	>1.65

Source: Adapted from Pal, S.C., Mukherjee, A., Pathak, S.R. 2003. *Cement and Concrete Research*, 33, 1481–1486; Smolczyk, H.G. 1980. Slag structure and identification of slag, principal report, *7th International Conference* on *Chemistry of Cement*, Vol. 1, pp. III/1–17, Paris.

Table 4.3 Chemical compositions and basicity indices of different granulated blast furnace slags

GBFS no	Oxide content (%)					Basicity indices of Table 4.2					
	SiO_2	CaO	MgO	Al_2O_3	Fe_2O_3	1	2	3	4	5	6
1	35.1	43.2	7.8	12.4	0.4	1.2	1.5	1.5	1.8	1.1	1.7
2	31.7	37.4	8.1	15.7	0.7	1.2	1.4	1.5	1.9	1.0	1.8
3	35.7	31.8	7.5	13.5	3.1	0.9	1.1	1.2	1.5	0.8	1.4
4	41.6	32.7	7.0	13.9	1.1	0.8	1.0	1.1	1.3	0.7	1.2
5	39.6	33.4	6.1	15.8	0.7	0.8	1.0	1.1	1.4	0.7	1.3

for five GBFS and given in Table 4.3. The last three slags in the table seem to be inferior to the first two. However, their performances in concrete were found satisfactory. It should be noted that glass content and fineness of a GBFS are equally or more significant than the chemical composition.

4.4 FINENESS

GBFS are used in cementitious systems after they are finely ground to a desired fineness which generally lies between 350 and 600 m²/kg (Blaine specific surface).

Chapter 5

Silica fume

5.1 GENERAL

Silica fume is a by-product of the silicon metal and alloy industries. Silicon metal and alloys are produced in electric arc furnaces in which quartz is reduced at temperatures above 1,700°C. Among the many chemical reactions that occur in the furnace, one of them results in the formation of SiO vapour which, upon oxidation and condensation, turns into very small spheres of amorphous silica, as shown in Figure 5.1.

In a typical ferrosilicon alloy plant, high-purity quartzite (containing about 95% SiO_2), scrap steel or iron or iron ore and coke, wood chips and coal are mixed in prescribed proportions and fed into the electric arc furnace. Reduced iron and silicon then combine to form the ferrosilicon alloy, which is taken from the bottom portion of the furnace into molds for cooling. Some of the SiO vapour escape from the furnace with stack gases and contacts with air to form the silica fume (SF) which is collected in baghouse filters. The flowchart for ferrosilicon alloy production is given in Figure 5.2. The amount of SF obtained is 10%–25% of the quartz fed into the furnace. Available global SF production is estimated as 1.2 million tons per year (Fidjestol and Dåstøl, 2014).

The chemical reaction that is intended in the production of silicon metal is

$$SiO_2 + 2C \rightarrow Si + 2CO \tag{5.1}$$

where SiO_2 is the quartz and carbon (C) comes from the coke, coal and wood chips.

An iron source is added in ferrosilicon alloy production. The intended reaction in this case becomes

$$SiO_2 + 2C + n\text{Fe} \rightarrow \text{Fe}_n\text{Si} + 2CO \tag{5.2}$$

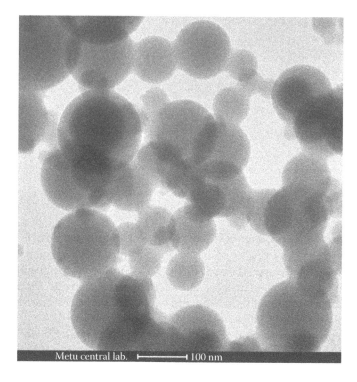

Metu central lab. |————| 100 nm

Figure 5.1 TEM image of SF particles.

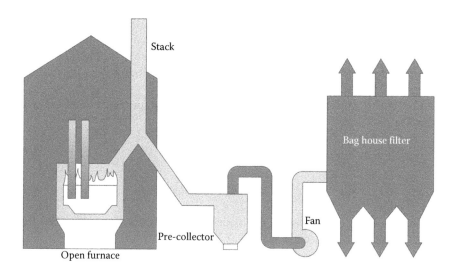

Figure 5.2 Simplified flow chart for SF production in a ferrosilicon alloy plant. (Adapted from Yeğinobalı, A. 2009. Silica fume. Its use in cement and concrete, TÇMB/ AR-GE/Y01.01, Turkish Cement Manufacturers' Association, 62pp. [in Turkish].)

However, the chemistry involved in the furnace is more complex than these two equations. SiO vapour and silicon carbide (SiC) form as intermediate products. SiO vapour that escapes the furnace contacts with air to form the SF

$$SiO + \frac{1}{2}O_2 \rightarrow SiO_2 \qquad (5.3)$$

There are other silicon alloys such as ferrochromium silicon, silicomanganese and calcium–silicon, the production of which also lead to SF. Chromium, manganese and lime are added to the raw materials to obtain the alloys, respectively. Obviously, the raw material composition also affects the SF composition.

Plants containing electric arc furnaces may have different designs; however, two basic types should be distinguished from the standpoint of SF production: those with and without heat recovery systems. The stack gases leave the furnace at about 800°C when it is equipped with a heat recovery system whereas their temperature is about 200°C when they leave the furnace without a heat recovery system. Most of the carbon is burnt in the former but the latter contains some unburnt carbon (Mehta, 1986a). These result in various differences in the SF obtained. On the other hand, unlike other industrial by-products used in cement and concrete, SF from a single source does not show significant variations in properties from one day to another (Pistilli et al., 1984).

5.2 CHEMICAL COMPOSITION

Examples for the chemical compositions of various SFs obtained from the production of silicon metal and different silicon alloys are given in Table 5.1. The chemical composition of the SF depends greatly on the composition of the principal product obtained from the furnace which in turn is directly related with the raw feed composition.

The silica content of the SF varies with that of the alloy produced. It is generally high (>80%) except for CaSi and SiMn alloys. SFs of these two alloy productions are not considered suitable for use in cement and concrete.

5.3 MINERALOGICAL COMPOSITION

Mineralogical analyses of different SFs by x-ray diffraction had shown that they are almost totally composed of amorphous silica. Typical x-ray diffractogram of SFs are given in Figure 5.3. As it can be clearly observed in

Table 5.1 Typical chemical compositions of silica fumes from different silicon metal and alloy plants

Component	Si	FeSi – 75%[a]	FeSi – 75% (with waste heat recovery)	FeSi – 50%	FeCrSi	CaSi	SiMn
			Amount (%) in SFs from different metal and alloy productions				
SiO_2	94.0	89.0	90.0	83.0	83.0	53.7	25.0
Fe_2O_3	0.03	0.6	2.9	2.5	1.0	0.7	1.8
Al_2O_3	0.06	0.4	1.0	2.5	2.5	0.9	2.5
CaO	0.5	0.2	0.1	0.8	0.8	23.2	4.0
MgO	1.1	1.7	0.2	3.0	7.0	3.3	2.7
Na_2O	0.04	0.2	0.9	0.3	1.0	0.6	2.0
K_2O	0.05	1.2	1.3	2.0	1.8	2.4	8.5
C	1.0	1.4	0.6	1.8	1.6	3.4	2.5
S	0.2		0.1				2.5
MnO		0.06		0.2	0.2		36.0
LOI	2.5	2.7		3.6	2.2	7.9	10.0

Source: Adapted from Aïtcin, P.C., Pinsonneault, P., Roy, D.M. 1984. *Bulletin of the American Ceramic Society*, 63, 1487–1491.

[a] 75% corresponds to the amount of silicon in the alloy.

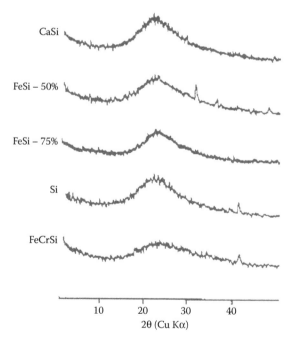

Figure 5.3 X-ray diffractograms of SFs. (From Aïtcin, P.C., Pinsonneault, P., Roy, D.M. 1984. *Bulletin of the American Ceramic Society*, 63, 1487–1491. With permission from American Ceramic Society Bulletin.)

the figure, the diffractograms shows halos corresponding to about 23° 2θ (Cu Kα) that indicates the amorphous siliceous character of SF.

5.4 FINENESS AND PARTICLE MORPHOLOGY

Some of the earlier investigations carried out by means of sieve analysis and Blaine air permeability method on the fineness of SF had found that approximately 3.7% of its particles retain on a 45 μm sieve and the specific surface area ranges between 3,000 and 8,000 m^2/kg (Pistilli et al., 1984). On the other hand, electron microscopy observations of SF have shown that it is composed of spherical particles of 10–300 nm size and the average particle size lies between 10 and 20 nm; and typical specific surface areas found by means of nitrogen adsorption method (BET) are around 20,000 m^2/kg (Mehta, 1986a).

Chapter 6

Limestone powder

6.1 GENERAL

Limestone has long been used in the construction industry as an aggregate for concrete and asphalt concrete. Its use in the cement industry as a raw material for PC is almost 200 years old. In powdered form, limestone has also been used for a long time in masonry cements for mortars and plasters. This use was standardised as early as 1932 (ASTM C 91, 2003). Incorporating limestone powder into PC dates back to the 1970s and currently its use in cements has become almost a common practice throughout the world (Erdoğdu, 2002). ASTM C 150 (2012) allows upto 5% limestone powder in PC whereas EN 197-1 (2012) specifies cements that may contain upto 35% limestone powder, besides using it as a minor additional constituent (<5%) in all types of common cements.

Limestone is sedimentary rock consisting primarily of calcite ($CaCO_3$). By definition, a rock that contains at least 50% (by mass) calcite is called limestone. Dolomite, quartz, feldspars, clay minerals, silt, chert, pyrite, siderite and other minerals may be present in small amounts. There are several varieties of limestone formed by different processes. It can be precipitated from water (inorganic limestone), secreted by marine organisms like algae and coral, or formed from the shells or skeletal parts of sea creatures (organic or biochemical limestone). Most limestone is layered and may contain fossils of shellfish and other animals that lived in shallow seas. Pure limestone is white or almost white in colour. However, it generally has shades of bluish grey, tan, yellow or brown depending on the type and amount of impurities present. The texture of limestone ranges from granular, coarsely crystalline to very fine grained, depending on the process of formation. SEM images of two types of limestone are shown in Figure 6.1.

6.2 CHEMICAL COMPOSITION

Examples for the chemical composition of limestone are given in Table 6.1. Since the major constituent of any limestone is calcium carbonate ($CaCO_3$),

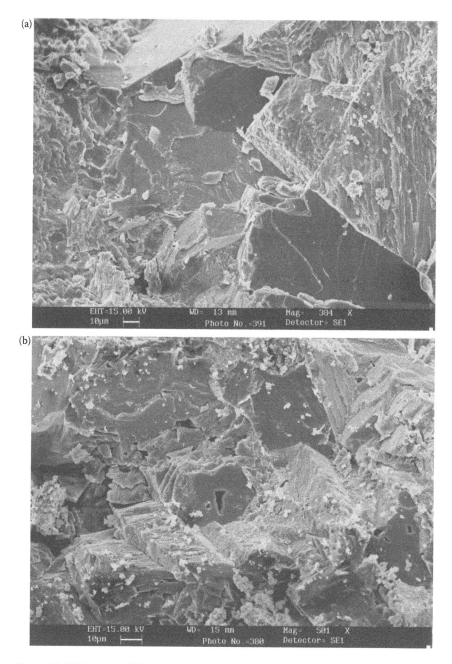

Figure 6.1 SEM images of (a) a limestone and (b) a chalky limestone. (Printed with permission from Erdoğdu, K. 2002. Hydration properties of limestone incorporated cementitious systems, PhD thesis, Middle East Technical University.)

Table 6.1 Typical chemical compositions of limestone used in cement industry

Component (%)	Limestone				
	A (Hawkins et al., 2003)	B (Erdoğdu, 2002)	C (TÇMB, 2014)	D (Hawkins et al., 2003)	E (Hawkins et al., 2003)
SiO_2	4.00	0.70	1.97	12.05	2.96
Fe_2O_3	0.77	0.07	0.23	3.19	0.79
Al_2O_3	0.30	0.11	0.68	1.22	0.30
CaO	51.40	54.90	53.24	43.05	52.30
MgO	1.30	0.38	1.07	1.68	1.30
SO_3	0.10	0.01	<0.01	0.56	0.03
Na_2O	0.01	0.02	0.09	0.12	0.04
K_2O	0.02	–	1.94	0.72	0.23
LoI	42.00	43.37	42.73	36.21	42.18

CaO content and loss on ignition (LoI) which is the weight loss upon heating to 900–1,000°C have the highest amounts in the chemical analyses. LoI is mainly the CO_2 given off upon the dissociation of calcium carbonate at 900°C (and that of magnesium carbonate impurity at about 600°C) as given in the equations below

$$CaCO_3 \xrightarrow{900°C} CaO + CO_2 \uparrow \tag{6.1}$$

$$MgCO_3 \xrightarrow{600°C} MgO + CO_2 \uparrow \tag{6.2}$$

Other oxides produced by the impurities present in the limestone are much less.

The suitability of a limestone as a cement constituent can be checked by three requirements: (1) Total $CaCO_3$ content ≥75.0%; (2) clay content, as determined by methylene blue adsorption test ≤1.2%; and (3) total organic carbon content ≤0.5% (all by mass) (EN 197-1, 2012).

6.3 MINERALOGICAL COMPOSITION

Limestone is basically calcite and sometimes calcite and aragonite (two polymorphic forms of $CaCO_3$). Dolomite ($MgCO_3$), quartz (SiO_2), clay minerals and feldspars may be present in much smaller amounts. Figure 6.2 shows the x-ray diffractograms of two types of limestone. The first one is almost totally calcite, whereas the second one contains dolomite and quartz as impurities.

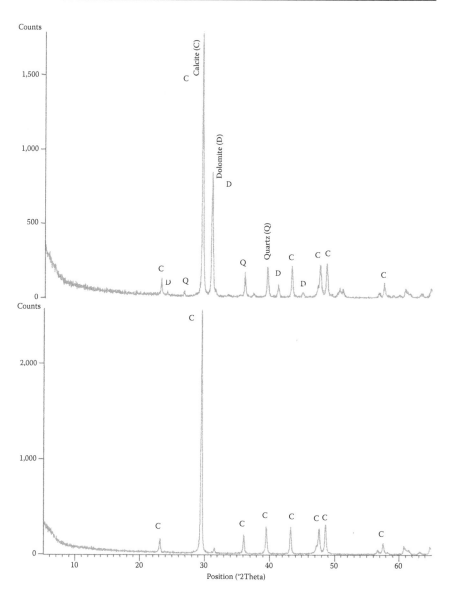

Figure 6.2 X-ray diffractograms of two types of limestone. (Courtesy of TÇMB R&D Institute, Ankara, Turkey.)

6.4 FINENESS AND PARTICLE SIZE DISTRIBUTION

Limestone is used as a mineral admixture or additive in powdered form. Being a comparatively softer material, limestone requires less energy to reach a specified fineness than a PC clinker, as illustrated in Figure 6.3. The desired

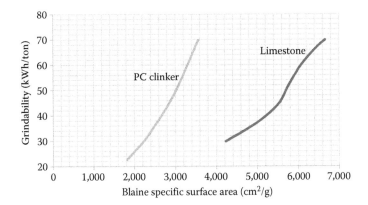

Figure 6.3 Comparison of the grindabilities of limestone and PC clinker in terms of energy required for a specified Blaine specific surface area. (Data from Erdoğdu, K. 2002. Hydration properties of limestone incorporated cementitious systems, PhD thesis, Middle East Technical University.)

fineness, in terms of specific surface area, generally lies between 350 and 600 m²/kg (Blaine). However, Blaine specific surface area by itself may not be sufficiently representative of the fineness of limestone powder due to several reasons: (1) Various types of limestone such as chalks and chalky limestone may contain some hygroscopic water which makes the ground material sticky and complicates further grinding operations by surrounding mill charges and plates; (2) Fine (≤15 μm) and coarse (≥90 μm) particles are relatively more than the medium size fractions in ground limestone and (3) Different size fractions of limestone have different grindabilities (Erdoğdu, 2002). Items (2) and (3) are graphically explained in Figure 6.4.

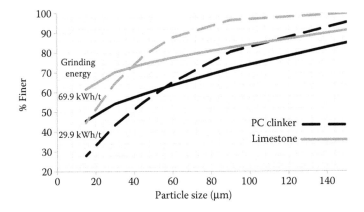

Figure 6.4 Particle size distributions of limestone and PC clinker ground at specified energies. (Data from Erdoğdu, K. 2002. Hydration properties of limestone incorporated cementitious systems, PhD thesis, Middle East Technical University.)

Chapter 7

Other mineral admixtures

There are many other less common mineral admixtures besides those described in the previous five chapters. The motive behind the considerable amount of research that has been carried out on the possibility of using these materials as components of cement and concrete is to find satisfactory solutions to the environmental issues related both with cement and concrete production and waste management. Some of these mineral admixtures like calcined clays, metakaolin (MK) and rice husk ash (RHA) are already in commercial use, whereas most others are under development or being investigated merely for scientific interest. It is neither possible nor meaningful to include all such materials within this book yet some general characteristics of those with pozzolanic properties are given in the following sections.

7.1 CALCINED SOILS

Calcined soils may be considered as the first artificial pozzolans as their use in ground form together with lime had been reported to date back to the Minoan and Mycenaean civilisations, more than 3600 years ago (Cook, 1986b; Theodoridou et al., 2013). There is also evidence that fragments of fired clay products were ground and mixed with lime to form a hydraulic binder as early as the Late Bronze Age in Cyprus (Theodoridou et al., 2013). The Greek, Roman, Indian and Ottoman civilisations were also aware of the cementitious value of the ground burnt clay mixed with lime. Crushed and ground bricks and tiles are still being used under the names like *horasan* in Turkey, *surkhi* in India and *homra* in Egypt in making mortars used mostly for the restoration of various historical structures.

Calcined clays may be used as a main ingredient in concrete according to ASTM C 618 (2012) and in CEM II, CEM IV and CEM V cements according to EN 197-1 (2012). However, their current use is not high due to the considerable competition they face from other mineral admixtures. Nevertheless, in places where materials like natural pozzolans, fly ash and granulated blast furnace slag are not available, calcined clays may be an economic and technical alternative.

7.1.1 Clays

Clay minerals are basically hydrous aluminium silicates having sheet-like or rod-like structure and very small particle size ($<2 \ \mu m$). Water present in clays may be adsorbed, interlayer or bound water. Various types of clays may contain significant amounts of magnesia, alkalies and iron oxide. Typical chemical compositions of various clay minerals are given in Table 7.1.

Clays are not pozzolanic materials. However, when they are calcined under controlled conditions, they may turn into a collapsed quasi-amorphous material which reacts with lime (Cook, 1986b). Heating the clay removes the adsorbed, interlayer and bound water, one after the other, as the temperature increases. The effect of temperature on the clay structure is usually monitored by differential thermal analysis (DTA). The endothermic peak obtained slightly above 600°C indicates the removal of bound water that leaves the clay an amorphous material. The exothermic peak at temperatures close to 1000°C, on the other hand, indicates recrystallisation.

The pozzolanic activity of calcined clay depends on its quasi-amorphous character which is obtained at an optimum temperature. The optimum calcination temperature is different for different clays. Below and above that temperature clay is not sufficiently reactive. It is necessary to note that the optimum calcination temperature is not the temperature corresponding to the main endothermic peak in the DTA curve but a temperature between the endothermic and exothermic peaks (Cook, 1986b; Sabir et al., 2001). The time which the clay is held at the optimum temperature is a significant parameter affecting the pozzolanicity of the calcined clays. Thus, an optimum calcination period also exists for different clays. Both the optimum-temperature and time of calcination are determined experimentally (Sabir et al., 2001; Rashad, 2013).

In a comparative study on the pozzolanic reactivity of different calcined clay minerals, kaolinite was determined to have higher potential for activation than illite and montmorillonite. Illite was found to behave almost like an inert filler and montmorillonite showed little pozzolanic activity in

Table 7.1 Various clay minerals and their typical chemical compositions

Clay mineral	Chemical formula	Composition, % by mass						
		SiO_2	Al_2O_3	Fe_2O_3	MgO	CaO	$K_2O + Na_2O$	H_2O
Kaolinite	$Al_2Si_2O_5(OH)_4$	46	40	–	–	–	–	14
Illite	$K_{0.6}(H_3O)_{0.4}Al_{1.3}Mg_{0.3}$ $Fe_{0.1}Si_{3.5}O_{10}(OH)_2(H_2O)$	54	17	2	3	1	7	12
Montmorillonite	$Na_{0.2}Ca_{0.1}Al_2Si_4O_{10}$ $(OH)_2(H_2O)_{10}$	43	19	–	–	1	2	35

30:70 (by mass) calcined clay—PC mixtures. On the other hand, calcined kaolinite substitution resulted in higher strength than the control PC mortar starting from 7 days (Fernandez et al., 2011).

The mineral admixture known as MK which is obtained by the calcination of kaolinite-rich clays over a temperature range of 650–800°C belongs to this group of materials. Although MK had received considerable scientific interest in the last 25 years, its first documented use dates back to the 1960s (Saad et al., 1982; Parande et al., 2009). Besides containing typically, 50%–55% SiO_2 and 40%–45% Al_2O_3 in amorphous state, its very fine particle size which is in the range of 0.5–5 µm, makes it a very reactive pozzolan. Its white colour is appropriate for architectural applications (Rashad, 2013). However, compared with another very reactive pozzolan, SF which is a by-product, MK requires considerable amount of energy for calcination.

7.1.2 Other soils

Shales which are sedimentary rocks obtained from clays under high overburden pressure have compositions similar to the clays from which they are derived. Therefore, the discussion made in the previous section also holds true for them.

Calcined lateritic and bauxitic soils which are rich in iron oxide and aluminium oxide, respectively, are stated to be reacting with lime although they do not contain an appreciable amount of silica (Cook, 1986b; Marwan et al., 1992; Péra and Momtazi, 1992).

7.2 RICE HUSK ASH

The artificial mineral admixtures discussed so far in the previous chapters were industrial by-products. However, RHA is of agricultural origin. Although there have been numerous attempts to use the ashes of other agricultural wastes as will be discussed in the subsequent sections, RHA is the one that has been investigated the most. This is not surprising since, according to 2013 figures, the total rice paddy production throughout the world has reached about 750 million tons/year and about 150 million tons of it is the husk (Aprianti et al., 2015).

Rice husk is the sheath that covers the rice grain and is removed during the refining process of rice. About 20% of the rice paddy is husk and about 20% of the husk is ash (Cook, 1986c; Aprianti et al., 2015). Like other agricultural residues, rice husk contains considerable amounts of organic constituents such as cellulose, lignin and fibres. Therefore, it is not usable in cementitious systems in its original state. On the other hand, when it is burnt, ash rich in silica remains. The chemical compositions of various RHAs used by different researchers are given in Table 7.2.

Table 7.2 Chemical compositions of RHA used in different studies

Constituent	(Cook, 1986c)	(James and Rao, 1986)	(Zhang and Malhotra, 1996)	(Nehdi et al., 2003)	(Zain et al., 2011)
	%, by mass				
SiO_2	92.15	94.43	87.20	94.60	86.49
Al_2O_3	0.41	–	0.15	0.30	0.01
Fe_2O_3	0.21	1.30	0.16	0.30	0.91
CaO	0.41	0.90	0.55	0.40	0.50
MgO	0.45	0.65	0.35	0.30	0.13
Na_2O	0.08	0.55	1.12	0.20	0.05
K_2O	2.31	1.32	3.68	1.30	2.70
LoI	2.77	3.00	8.55	1.80	8.45

The combustion process of the rice husk first removes the absorbed moisture (~100°C), then the volatile matter (~350°C) and the carbon (~450°C) are removed. The silica in the husk is in completely amorphous form upto 600°C. As the temperature increases, silica undergoes various structural transformations. When the temperature exceeds 600°C, first the quartz and then with the increasing temperature, other crystalline forms of silica (crystobalite and tridymite) may be detected. Burning at temperatures above 800°C results in almost completely crystalline silica (Cook, 1986c). RHA with good pozzolanic activity is produced at burning temperatures between 500°C and –700°C. Although very different burning periods have been suggested so far, 12 h incineration of rice husk at 500°C is proposed for obtaining highly reactive RHA (Nair et al., 2008).

The RHA is usually ground to an appropriate fineness. The grinding time is related with the nature of the silica in the ash. As the amount of crystalline silica increases, the grinding period to achieve a specified fineness also increases. Generally, a 30–120 min grinding period is sufficient in ball mills to obtain more than 90% of the ash passing a 45 μm sieve (Cook, 1986c; Zain et al., 2011). On the other hand, the irregularly shaped particles of RHA has a cellular structure that leads to excessive water absorption which may result in workability problems in cementitious systems. Therefore, in order to break down the cellular structure and reduce the amount of micropores in the ash, grinding periods above 60 min are recommended (Zain et al., 2011).

7.3 OTHER INDUSTRIAL AND AGRICULTURAL WASTES AS POSSIBLE MINERAL ADMIXTURES

Population growth rates, changing consumption patterns, increasing production and faster consumption of resources throughout the world have resulted in a continuously growing amount of wastes. Wastes may be generated during the raw material extraction, production and consumption processes. Although it is difficult to define waste because the legal, economic and environmentalist approaches usually differ from each other, recyclable or reusable disposed materials are generally not considered as waste. Increasing awareness for sustainability and legal, economic and ecological restrictions have led to many research and investigations on the recycling and reuse of wastes.

Cement and concrete industries are considered as safe and economic means of using many wastes. The main reason for this lies in the fact that several industrial and agricultural by-products such as FA, BFS, SF and RHA which were once thought to be wastes have been successfully used in cement and concrete production in amounts that are not to be underestimated. The three-fold (ecological, economical and technical) benefits obtained by using these materials encouraged many researchers to investigate the possibilities of using other wastes in cement and concrete manufacturing.

Many different ways of using wastes in the cement and concrete industry were experienced and are being investigated. Some wastes are being used as secondary fuels and raw materials in cement production and some are being used as aggregates or fillers for special types of concrete. Some of the wastes having the potential of being cement or concrete mineral admixtures will be covered within this context.

7.3.1 Waste glass

The global quantity of waste glass is estimated to be about 14 million tons per year (Jani and Hogland, 2014). Its high silica content (>70%) and amorphous nature makes it a possible mineral admixture for cement and concrete. Indeed, numerous studies as reviewed by Shi and Zheng (2007), Federico and Chidiac (2009) and Jani and Hogland (2014) had been carried out to reveal that when finely ground with particle sizes less than 40 μm, waste glass shows satisfactory pozzolanic properties. The major problem of using waste glass in cementitious systems is the risk of alkali–silica reactivity. However, it was reported that partial cement replacement by finely ground waste glass results in improved resistance to alkali–silica reaction (Matos and Sousa-Coutinho, 2012).

7.3.2 Brick and tile rejects

Finely ground clay bricks obtained from the rejects of clay brick and tile manufacturing or demolished masonry may be used as a mineral admixture in concrete. Crushed and ground fired clay products have long been used with lime to form a hydraulic binder, as previously stated. Recent research had shown that ground brick may be used efficiently as a pozzolanic material to improve the resistance of concrete to sulphate attack and alkali–silica expansion (O'Farrell et al., 1999; Turanlı et al., 2003; Bektaş et al., 2008; Bektaş and Wang, 2012; Pereira-de-Oliveira et al., 2012).

7.3.3 Stone wastes

The powdered substances obtained as the crushing or cutting wastes of various stone industries were attempted to be used as mineral admixtures in concrete. Limestone (Felekoğlu, 2007), calcite (Temiz and Kantarcı, 2014), marble (Corinaldesi et al., 2010), granite (Vijayalakshmi et al., 2013) and basalt dusts (Laibao et al., 2013) were reported to result in satisfactory concrete properties when they are used to replace fair amounts (~10%, by mass) of PC. The major reason for using such stone powders in concrete, besides their elimination, is the improvement that they may bring to the cohesion of fresh concrete.

7.3.4 Municipal solid waste ash

Incineration had become a common way of eliminating municipal solid wastes (MSW). There are two different incineration facilities: the refuse-derived fuel process uses presorted MSW in which metals and glass are removed and the mass-burn process uses MSW as received. Both processes reduce the amount of MSW and supply energy. Incineration results in 1%–30% ash by weight of the wet MSW. About 80% of the ash is bottom ash and the rest is FA. Sometimes these ashes are mixed to form combined ash (Ferreira et al., 2003; Siddique, 2010). Both ashes may contain serious amounts of leachable heavy metals but total heavy metal concentration is generally higher in the fly ashes. Chemical and elemental compositions of various bottom ashes and fly ashes of MSWs are given in Table 7.3. MSW ashes contain more than 60% amorphous phase (Filipponi et al., 2003).

7.3.5 Agricultural wastes

There have been many attempts to utilise agricultural residues in construction products like insulation boards, particle boards, blocks, etc. However, they have found limited commercial applications. Moreover, most of the agricultural residues are not considered as waste since they can be used as stock feed or fuel (Cook, 1986c). If they are used as fuel, their ashes may provide pozzolanic materials suitable for use in cement and concrete. Besides the rice husk which was discussed previously, the residues of oil palms, corn cobs, bagasse and bamboo leaves are other materials that were investigated recently for their pozzolanic potentials.

Oil palms are tropical trees from which palm oil is obtained. Indonesia, Malaysia and Thailand are the leading palm oil producers in the world.

Table 7.3 Oxide and elemental compositions of some MSW ashes (based on dry weight)

	Oxide composition (% by dry weight)					Heavy metals (mg/kg of dry weight)			
Oxide	BA 1 (Filipponi et al., 2003)	BA 2 (Filipponi et al., 2003)	FA 1 (Rémond et al., 2002)	FA2[a] (Aubert et al., 2004)	Heavy Metal	BA 1 (Filipponi et al., 2003)	BA 2 (Filipponi et al., 2003)	FA 1 (Rémond et al., 2002)	FA2[a] (Aubert et al., 2004)
SiO_2	47.76	56.99	27.23	20.67	Zn	1,443	21,344	11,000	24,046
Al_2O_3	10.55	9.20	11.72	10.01	Pb	1,287	1,473	4,000	8,816
Fe_2O_3	8.61	3.97	1.80	2.73	Cu	2,660	1,526	670	1,714
CaO	16.45	13.22	16.42	25.23	Cr	79	80	450	2,078
MgO	3.67	3.46	2.52	2.74	Cd	<8	<8	270	586
K_2O	1.41	1.35	5.80	1.35	Sn	nd	nd	180	2,883
Na_2O	3.51	5.87	5.86	1.35	Sb	nd	nd	110	1,457
MnO	0.13	0.08	0.05	0.20	Se	nd	nd	50	nd
TiO_2	0.79	0.49	0.84	1.73	V	nd	nd	32	30
P_2O_5	1.29	0.70	0.34	13.56	Mo	nd	nd	25	49
SrO	nd	nd	0.01	nd	As	nd	nd	21	120
Lol	3.47	2.94	13.00	6.50	Co	nd	nd	21	79

[a] This fly ash was treated to recover sodium, stabilise the heavy metals and eliminate organic matter.

Table 7.4   Typical chemical compositions of various agricultural residues

Oxide (% by mass)	Oil palm ash	Corn cob ash	Bagasse ash	Bamboo leaf ash
SiO_2	59.6–66.9	65.4–67.3	60–65.3	80.4
Al_2O_3	2.5–6.4	6.0–9.1	4.7–9.1	1.22
Fe_2O_3	1.9–5.7	3.8–5.6	3.1–5.5	0.71
CaO	4.9–6.4	10.3–12.9	4.0–10.5	5.06
MgO	3.0–4.5	1.8–2.3	1.1–2.9	0.99
SO_3	0.3–1.3	1.0–1.1	0.1–0.2	1.07
Na_2O	0.2–0.8	0.4–0.5	0.3–0.9	0.08
K_2O	5.0–7.5	4.2–5.7	1.4–2.0	1.33
LoI	6.6–10.0	0.9–1.5	15.3–19.6	8.04

Source: Reprinted from *Construction and Building Materials*, 74, Aprianti, E. et al.,
Supplementary cementitious materials origin from agricultural wastes: A review,
176–187, Copyright 2015, with permission from Elsevier.

The waste after palm oil production constitutes of shells and fibres which are burnt at 800–1,000°C in biomass thermal power plants. About 5% (by mass) of the oil palm residue is ash which is rich in silica (Aprianti et al., 2015).

Corn cob is the waste of maize production. United States, South Africa and Nigeria are the largest maize producers in the world. Corn cob when burnt at temperatures around 700°C, results in about 12% ash which is rich in silica and with a fair amount of CaO (Adesanya and Raheem, 2009).

Bagasse is the residue of the raw sugar production process from sugar cane. It is then used in cogeneration and combustion processes to result in about 14% ash. Countries such as India, Thailand, Brazil, Columbia, Phillipines, Indonesia and Malaysia produce a considerable amount of bagasse ash. The silica content of bagasse ash is at least 60% (Aprianti et al., 2015).

Bamboo leaf ash was stated to be a promising pozzolanic material with silica content more than 80% (Cocina et al., 2011).

The chemical compositions of these ashes are given in Table 7.4. There may be many other agricultural residues the ashes of which possess pozzolanic properties. However, if such residues can be used economically otherwise for animal feeding or as raw materials for certain other manufacturing processes (like wood wastes) they may not be feasible for use in concrete.

Chapter 8

Effects of mineral admixtures on hydration of portland cement

Hydraulic binders are powdered substances that result in a firm solid upon reacting with water which then remains stable in aqueous environments. Although there are different varieties of inorganic hydraulic binders such as calcium sulphoaluminate, calcium aluminate, phosphate, portland and blended (or composite) cements, this chapter deals with the hydration of portland and blended cements.

Portland cement is defined as 'hydraulic cement produced by pulverising PC clinker, and usually containing calcium sulfate' (ASTM C 219, 2001). Hydraulic binders that contain latent hydraulic materials, pozzolans or non-reactive constituents together with PC clinker (and usually with gypsum) are called composite or blended cements.

8.1 GENERAL ABOUT PCs

PC clinker (Figure 8.1) is a product of burning a raw mix containing appropriate amounts of CaO, SiO_2, Al_2O_3, Fe_2O_3 (plus other oxides of much smaller amounts) to 1,400–1,500°C. $CaCO_3$ in the calcareous component of the raw mix reveals CaO on dissociation at around 900°C and CaO reacts with the SiO_2, Al_2O_3 and Fe_2O_3 coming from the argillaceous component of the raw mix to yield clinker minerals. The four major minerals (Figure 8.2) present in the clinker are tricalcium silicate (Ca_3SiO_5; also called as alite), dicalcium silicate (Ca_2SiO_4; also called as belite), tricalcium aluminate ($Ca_3Al_2O_6$) and calcium aluminate ferrite (a solid solution of calcium aluminate ferrite compounds of varying alumina–iron oxide ratios; also called the ferrite phase). The ferrite phase is sometimes called tetracalcium alumino ferrite ($Ca_4Al_2Fe_2O_{10}$) which is considered as an approximate composition of the solid solution of calcium aluminate ferrite compounds in PC. In the industrial PCs, these minerals are not present as pure compounds but contain variable amounts of foreign ions in their crystal lattices.

The clinker is then cooled and ground usually with 3%–5% (by mass) gypsum to obtain PC (Figure 8.3).

Figure 8.1 PC clinker.

The compounds present in cements are generally expressed as the sum of their oxides although they do not have a separate existence within the compounds. For example, tricalcium silicate (Ca_3SiO_5) can be written as $3CaO \cdot SiO_2$. Therefore, it is a common practice to determine the chemical compositions of inorganic binders by means of oxide analysis. Chemical compositions of the phases of cements are usually given by abbreviated formulas in which the oxides are expressed as one-letter notations which are listed in Table 8.1. Accordingly, tricalcium silicate, dicalcium silicate and tricalcium aluminate may be expressed as C_3S, C_2S and C_3A, respectively, in terms of cement chemistry abbreviations. Many other compounds that form on the reaction of cements with water may have more sophisticated

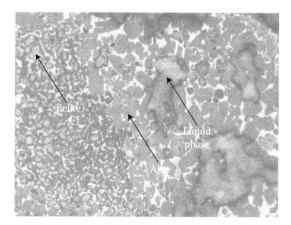

Figure 8.2 PC clinker minerals. (Courtesy TÇMB R&D Institute, Ankara, Turkey.)

Figure 8.3 Cement particles. (Courtesy TÇMB R&D Institute, Ankara, Turkey.)

standard formulas which can be expressed more conveniently by using cement chemistry abbreviations. For example, tricalcium aluminate trisulphate hydrate, a compound which is more commonly known as ettringite, is a reaction product of C_3A, gypsum and water and has the standard formula $Ca_6Al_2S_3O_{50}H_{64}$. When the standard formula is separated into its oxides, it can be written as $6CaO \cdot Al_2O_3 \cdot 3SO_3 \cdot 32H_2O$ which then can be abbreviated as $C_6A\bar{S}_3H_{32}$.

Table 8.1 Cement chemistry abbreviations of oxides in cements

Oxide	Standard formula	Cement chemistry notation
Aluminium oxide	Al_2O_3	A
Calcium oxide	CaO	C
Carbon dioxide	CO_2	$\bar{C}$
Iron oxide	Fe_2O_3	F
Water	H_2O	H
Potassium oxide	K_2O	K
Magnesium oxide	MgO	M
Sodium oxide	Na_2O	N
Phosphorus pentoxide	P_2O_5	P
Silicon dioxide	SiO_2	S
Sulphur trioxide	SO_3	$\bar{S}$
Titanium oxide	TiO_2	T

Table 8.2 Typical compound composition of an ordinary PC

Name of the compound	Notation	Amount (%, by mass)
Tricalcium silicate	C_3S	50.0
Dicalcium silicate	C_2S	25.0
Tricalcium aluminate	C_3A	12.0
Ferrite phase	C_4AF	8.0
Gypsum	$C\bar{S}H_2$	4.0

The compound composition of a typical ordinary PC is given in Table 8.2. The two calcium silicates account for about 75% (by mass) of PCs. It should be noted that the total amount of the compounds is not equal to 100%. The missing amount is due to the presence of some minor constituents and impurities.

8.2 PC HYDRATION

The constituents of PC undergo a series of chemical reactions when mixed with water. The overall process is named as hydration. Chemically speaking, hydration is the general term for the reaction of an anhydrous material with water, yielding the reaction products which are called hydrates (Odler, 1988). It is necessary to note that PC is a multi-component system therefore its hydration is a rather complex process involving numerous simultaneous and successive reactions. Besides the four major clinker compounds, free calcium oxide, calcium sulphate and sodium and potassium sulphates also participate in the hydration process.

The explanation of PC hydration is generally done by first considering the individual hydrations of the major compounds although their interactions with each other and with other phases present in the cement affect both the mechanism and the kinetics of hydration.

8.2.1 Hydration of tricalcium silicate

Tricalcium silicate (C_3S) is an essential constituent of PC. Its crystalline lattice contains calcium cations, silicate and oxygen anions. C_3S is not pure in commercial PCs but may contain aluminium, magnesium, sodium, potassium and iron as foreign ions. Therefore, it is generally called *alite* due to its resemblance to a naturally occurring mineral of the same name.

Generally, a direct correlation is made between the hydration of PC and that of C_3S as it is the compound of largest amount in most PCs. C_3S hydration results in two products: calcium silicate hydrate (C–S–H) gel and calcium hydroxide (CH; also called portlandite). The stoichiometric equation for full C_3S hydration is written as

$$2C_3S + 6H \rightarrow C_3S_2H_3 + 3CH \tag{8.1}$$

$C_3S_2H_3$ is an approximate composition of the C–S–H on full hydration of C_3S. C–S–H composition changes considerably during the course of hydration with the amount of water used, temperature and degree of hydration. Lime–silica ratio (C/S) upon full hydration is 1.50 and it never exceeds 2.00, at any stage of hydration. Since the molar ratio of lime-to-silica of C–S–H is always lower than that of the original C_3S, a considerable amount of CH forms. For the case of complete hydration, 39% of the total hydration products of C_3S is CH and 61% is C–S–H.

The kinetics of C_3S hydration is studied by determining the amount of remaining unhydrated C_3S with time by quantitative x-ray diffraction. However, this method is not found sufficiently precise for the early period of hydration where the degree of hydration is less than 0.1 (Odler, 1988). Indirect methods such as determination of heat evolution and/or rate of heat evolution and estimating the amount of chemically bound water are also being used in the study of hydration kinetics. Typical time-degree of hydration relations for C_3S are shown in Figure 8.4. It is possible to

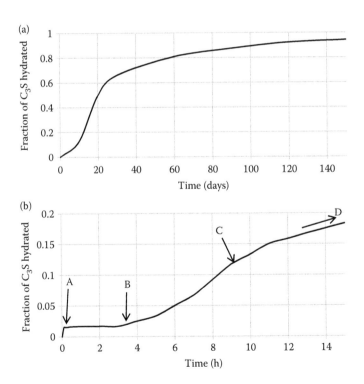

Figure 8.4 Typical (a) long-term and (b) short-term progress of hydration of C_3S. (Reprinted from Hydration, setting and hardening of portland cement, in *Lea's Chemistry of Cement and Concrete*, 4th Ed. [Ed. P.C. Hewlett], Odler, I, pp. 241–289, Elsevier, London, Copyright 1988, with permission from Elsevier.)

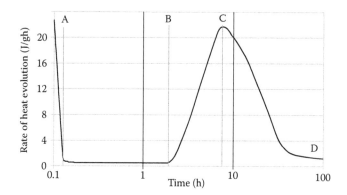

Figure 8.5 Rate of heat evolution during C₃S hydration. (Adapted from Mindess, S., Young, J.F. 1981. *Concrete*, Prentice-Hall, Englewood Cliffs, NJ.)

distinguish different stages of C₃S hydration from Figure 8.4: 0–A: pre-induction period; A–B: induction (dormant) period; B–C: acceleration period and C–D: deceleration period. On the other hand, this sequence of hydration stages is more conveniently described by referring to a calorimetric curve as illustrated in Figure 8.5. Since C₃S hydration, as well as all hydration reactions of PCs, is exothermic, the rate of heat evolution during the course of hydration shows the different stages more clearly.

During the pre-induction period which lasts only a few minutes, a very high rate of hydration of C₃S occurs and rapidly slows down to almost zero when the induction period starts. On contact with water oxygen ions in the C₃S are released into the surrounding liquid phase to form hydroxyl ions as a result of protonation:

$$O^{2-} + H^+ \rightarrow OH^- \tag{8.2}$$

Similarly, the silicate ions of C₃S enter into solution to form hydrogen silicate ions:

$$SiO_4^{4-} + nH^+ \rightarrow H_nSiO_4^{(4-n)} \tag{8.3}$$

These combine with the Ca²⁺ ions to form C–S–H which precipitates. Thus, a layer of C–S–H is formed on the surface of the C₃S particles which act as a barrier to the further contact of water with the underlying C₃S and further release of hydroxyl, silicate and calcium ions to the surrounding liquid within the induction period (Odler, 1988). Another theory on the formation of this immediate C–S–H states on the release of Ca²⁺ ions from the C₃S particles results in a silica-rich surface layer which then adsorbs the calcium ions in the surrounding liquid and thus a thin and almost impervious C–S–H layer covers the C₃S particles (Odler, 1988).

The induction period lasts several hours. The reason for the ending of the induction period and start of the acceleration period are not obvious. There are several different theories developed for explaining the phenomena. Without going into the details of the theories, it can be stated that the initially formed C–S–H layer undergoes various changes in composition and/or morphology with time which makes it more permeable or it breaks down due to the osmotic pressure developed in the liquid at the interface of C–S–H layer and the underlying C_3S. Thus, further hydration becomes possible.

The rate of hydration in the acceleration period increases rapidly and reaches a maximum within 5–10 h. CH concentration of the liquid phase becomes maximum within this period and it consequently starts to precipitate.

Hydration of C_3S starts to slow down after reaching the maximum rate. It continues, although at a reducing rate, as long as favourable curing conditions (humidity and temperature) are attained.

8.2.2 Hydration of dicalcium silicate

There are five possible polymorphic forms (α, α'_L α'_H β and γ) of dicalcium silicate (C_2S) which form by solid state reactions of CaO and SiO_2. γ-C_2S is obtained if CaO and SiO_2 are pure. However, this polymorph is not cementitious. α'_L, α'_H, and α-C_2S forms are obtained at 860°, 1160° and 1420°C, respectively. The reactions leading to the formation of these are reversible and upon cooling, at around 650°C, β-C_2S forms. β-C_2S is metastable unless there are impurity ions in its crystal lattice and converts into γ-C_2S. Among these five polymorphs, β-C_2S is the only form that is sufficiently reactive with water (Glasser, 1988). Fortunately, it is a regular constituent of commercial PCs.

C_2S hydrates in a similar manner to C_3S, but with a much slower rate, as illustrated in Figure 8.6. The stoichiometric equation for full C_2S hydration is written as

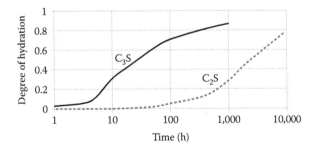

Figure 8.6 Comparison of C_2S and C_3S hydration rates. (Adapted from Odler, I. 1988. Hydration, setting and hardening of portland cement, in *Lea's Chemistry of Cement and Concrete*, 4th Ed. [Ed. P.C. Hewlett], pp. 241–289, Elsevier, London.)

$$2C_2S + 4H \rightarrow C_3S_2H_3 + CH \tag{8.4}$$

The hydration products of both calcium silicates are the same but their relative amounts are different. The amount of CH produced on complete hydration of C_2S (about 18% of total hydration products) is much less than that produced by C_3S (about 39% of total hydration products). Although the process is much slower, the mechanism of C_2S hydration is similar to that of C_3S. The hydration of C_2S starts with an exothermic peak as C_3S. However, further hydration proceeds gradually since C_2S is less reactive than C_3S and much less heat is liberated by the hydration of C_2S. The second peak that is observed in the rate of heat of hydration curve of C_3S is barely detectable for C_2S (Mindess and Young, 1981; Odler, 1988).

8.2.3 Hydration of tricalcium aluminate

Tricalcium aluminate (C_3A) usually exists as cubic crystals. However, if the amount of alkalies $(Na_2O$ and $K_2O)$ is high in the PC, orthorhombic or monoclinic forms may also be observed. Although the rates of reaction may change, the hydration of all three forms of C_3A are similar. C_3A hydration leads to two products:

$$2C_3A + 27H \rightarrow C_2AH_8 + C_4AH_{19} \tag{8.5}$$

Neither of these hydration products is stable and they convert into C_3AH_6, which is the only stable form of calcium aluminate hydrate at ordinary temperatures (Mindess and Young, 1981; Tumidaski and Thomson, 1994):

$$C_2AH_8 + C_4AH_{19} \rightarrow 2C_3AH_6 + 15H \tag{8.6}$$

C_3A hydration occurs very rapidly on mixing with water and slows down due to the formation of C_2AH_8 and C_4AH_{19} layer on the C_3A particles. It is then reaccelerated when these hydration products turn into C_3AH_6 (Boikova et al., 1980; Plowman and Cabrera, 1984; Pommersheim and Chang, 1986; Tong et al., 1991).

The hydration of C_3A alone results in a very rapid stiffening (flash setting) with very high heat evolution within few minutes after mixing with water. In the PCs, a small amount of calcium sulphate (generally as gypsum) is always present in order to slow down the hydration of C_3A, so that workability can be attained for a reasonable period of time. $C_6A\bar{S}_3H_{32}$ (ettringite) is obtained when C_3A + calcium sulphate system hydrates:

$$C_3A + 3C\bar{S}H_2 + 26H \rightarrow C_6A\bar{S}_3H_{32} \tag{8.7}$$

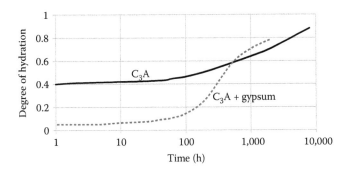

Figure 8.7 Comparison of C_3A and C_3A + gypsum hydration rates. (Adapted from Odler, I. 1988. Hydration, setting and hardening of portland cement, in *Lea's Chemistry of Cement and Concrete*, 4th Ed. [Ed. P.C. Hewlett], pp. 241–289, Elsevier, London.)

Ettringite is an expansive compound and if it forms under confined conditions (e.g. if the paste is hardened), it leads to the disruption of the system. Therefore, the amount of gypsum added as a retarder in the PC is so adjusted that most of C_3A is consumed to form ettringite while the paste is still plastic. Thus, the expansive compound formed does not cause disruptive internal stresses within the body. If the calcium sulphate is consumed before all the C_3A has hydrated, the previously formed ettringite reacts with the remaining C_3A and water to convert into monosulphoaluminate hydrate ($C_4A\bar{S}_{12}$):

$$C_6A\bar{S}_3H_{32} + 2C_3A + 4H \rightarrow 3C_4A\bar{S}H_{12} \qquad (8.8)$$

When monosulphoaluminate hydrate comes into contact with a new source of sulphate ions, ettringite forms again by a process known as sulphate attack, which is one of the major durability problems encountered in cementitious systems.

Typical kinetics of hydration of C_3A alone and C_3A + gypsum are shown in Figure 8.7. It should be noted that the initial amount of C_3A hydrated in the presence of calcium sulphate is considerably reduced when compared to that of C_3A alone.

Although the reactions are completely different, the calorimetric curve of C_3A + gypsum hydration shown in Figure 8.8 looks like that of C_3S given in Figure 8.5. However, the heat evolution is much greater in the case of C_3A + gypsum (Mindess and Young, 1981).

8.2.4 Hydration of ferrite phase

Calcium aluminoferrite phase in the PCs consists of a series of compounds whose compositions vary from $C_2(A_{0.7}, F_{0.3})$ to $C_2(A_{0.3}, F_{0.7})$ forming a solid

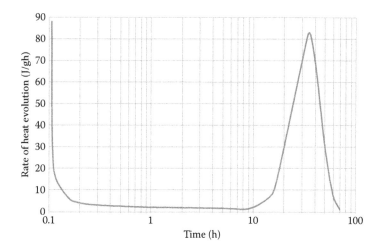

Figure 8.8 Rate of heat evolution during C_3A + gypsum hydration. (Adapted from Mindess, S., Young, J.F. 1981. *Concrete*, Prentice-Hall, Englewood Cliffs, NJ.)

solution. For a long time, it was named tetracalcium aluminoferrite (C_4AF), although such a compound does not exist. Later, abbreviations like $C_2(A,F)$ or F_{ss} became more common.

Hydration products formed by the ferrite phase are similar to those formed from C_3A. Fe^{3+} ions replace Al^{3+} ions to a limited degree in the crystal lattices of the hydration products. Rates of pure $C_2(A,F)$ and pure C_3A are comparable. However, the rate of hydration of the ferrite phase becomes slower as its alumina–iron oxide ratio decreases (Collepardi et al., 1979; Negro and Stafferi, 1979). The hydration rate is slowed down in the presence of gypsum, as in the case of C_3A hydration (Negro and Stafferi, 1979; Fukuhara et al., 1981).

8.2.5 Hydration of PC

Hydration of PC is affected by a number of factors such as (Odler, 1988)

- Compound composition of the cement
- Amount of gypsum present in the cement
- Presence of foreign ions in the individual compounds
- Fineness of the cement and its particle size distribution
- Water–cement ratio of the mixture
- Ambient temperature
- Presence and amount of admixtures and/or additives

The anhydrous cement compounds start to dissolve when they come into contact with water and form hydrated products of much lower solubility which precipitate out of the solution to deposit into previously water-filled spaces (Mehta, 1983). There are two theories proposed to describe the mechanism of such a process: (1) through-solution hydration and (2) topochemical hydration (Osbaeck and Jons, 1980). If the hydration rate is controlled by the precipitation of hydration products, water becomes oversaturated and precipitation randomly occurs within the whole water volume. This is known as through-solution hydration. On the other hand, if the rate of hydration is controlled by dissolution of the starting compounds, precipitation of hydration products is confined to the surface of the solid particles. The process is called topochemical hydration (Odler, 2000). The through-solution process dominates the early stages of PC hydration and at later ages the topochemical process becomes more pronounced (Mehta and Monteiro, 2006).

The hydration of PC occurs in four stages, similar to those of C_3S: (1) pre-induction period, (2) induction (dormant) period, (3) acceleration period and (4) deceleration period. It should be kept in mind that PC hydration is much more complex than those of the individual phases since their interactions with each other and with other minor phases such as free CaO, MgO, alkali sulphates, etc. may alter both the mechanism and the kinetics of hydration. PC hydration is summarised in Table 8.3. The typical progress of hydration in terms of hydration products formed is shown in Figure 8.9.

8.3 INFLUENCE OF MINERAL ADMIXTURES ON HYDRATION

The mechanism and kinetics of the hydration of mineral admixture incorporated cements is of great importance. However, the accumulation of knowledge on these aspects is still at an infant stage and of a broad nature (Scrivener and Nonat, 2011). Mineral admixtures are grouped into three categories as (1) materials of low or no reactivity, (2) pozzolanic materials and (3) latent hydraulic materials. Obviously, the rates of reaction and their interactions with the PC phases are of significant chemical nature for the latter two categories. The materials in the first category are considered as 'inert' substances and their effect on hydration is more of a physical nature. This however, does not mean that materials belonging to the first group are totally chemically inert and materials belonging to the second and third categories do not have any physical effect on hydration. Therefore, mineral admixtures modify the hydration process by both physical and chemical means which are sometimes very difficult to dissociate (Lawrence et al., 2003). Furthermore, due to the differences in their chemical and mineralogical compositions and physical properties, the extent of the influence of different mineral admixtures on hydration may vary considerably.

Table 8.3 Summarised mechanism and kinetics of PC hydration

Stage	Mechanism and kinetics
Pre-induction period	• Alkali sulphates dissolve immediately, releasing K^+, Na^+, and SO_4^{2-}. • Gypsum dissolves and releases Ca^{2+} and SO_4^{2-}. • C_3S starts to dissolve and a layer of C–S–H precipitates on the cement particles. Due to the lower lime–silica ratio of the C–S–H than C_3S, additional Ca^{2+} and OH^- are released into the liquid phase. About 2%–10% of C_3S is hydrated in this period. • C_3A dissolves and reacts with Ca^{2+} and SO_4^{2-} ions in the liquid phase resulting in ettringite that also precipitates on the surface of the cement particles. About 5%–25% C_3A hydrates within this period. • Ferrite phase hydrates in a similar manner as C_3A, yielding iron-bearing ettringite. • Pre-induction period has the highest rate of hydration but lasts only a few minutes.
Induction (dormant) period	• The hydration of all mineral phases proceeds extremely slowly. The rate of hydration is slowed down significantly. • The layer that forms on the surface of the cement particles upon the precipitation of initial calcium silicate hydrate and calcium alumino hydrates during the pre-induction period acts as a barrier to further hydration. • CH concentration in the liquid phase reaches its maximum and starts to reduce. • Dormant period takes a few hours.
Acceleration period	• The rate of C_3S hydration increases significantly. Higher SO_3 contents of the cement cause more and higher C_2S–C_3S ratios cause less acceleration (Osbaeck and Jons, 1980). • C_2S hydration becomes noticeable. • C_3A hydration is also accelerated for a few hours. • Ferrite phase hydrates in a similar manner to C_3A but at a slower rate. • CH precipitates from the liquid phase resulting in significant reduction of Ca^{2+} concentration of the liquid phase. • SO_4^{2-} concentration of the liquid phase decreases as most of it is used to produce ettringite and some is adsorbed by C–S–H. • Acceleration period may take 10–20 h.
Deceleration period	• Hydration rate slows down as the amount of unreacted cement gets less. • C–S–H continues to be formed as both C_3S and C_2S continue to hydrate. Contribution of C_2S in C–S–H formation is more at this stage. • Consequently, rate of CH formation is reduced. • Previously formed ettringite may react with remaining unreacted C_3A to yield calcium monosulphate hydrate. • Hydration process continues till all the cement is hydrated or all the water is consumed.

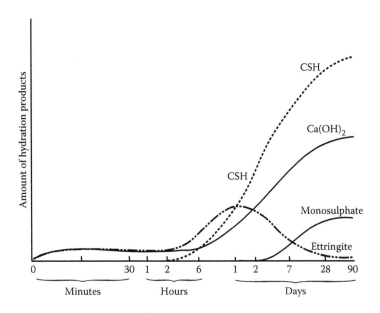

Figure 8.9 Rates of hydration product formation in PC.

Most of the research carried out on the effect of mineral admixtures on hydration so far report that they enhance cement hydration (Soroka and Stern, 1976; Takemoto and Uchikawa, 1980; Meland, 1983; Buil et al., 1984; Cheng-yi and Feldman, 1985; Gutteridge and Dalziel, 1990a, 1990b; Maltais and Marchand, 1997) whereas some report retardation (Carette and Malhotra, 1983; Meland, 1983; Fajun et al., 1985; Von Berg and Kukko, 1991; Tarun and Singh, 1997), the possible reasons for which are discussed later.

8.3.1 Physical influence

The physical effects of mineral admixtures on hydration may be considered from four aspects: (1) cement dilution effect, (2) dispersion effect, (3) modification of particle size distribution and (4) nucleation effect. The first one is encountered when the mineral admixture is used as a partial cement replacement material. An increased amount of mineral admixture results in a decreased amount of cement and naturally less cement means less hydrated material. This effect is shown in terms of heat of hydration at 2, 7 and 28 days in Figure 8.10 for ground granulated blast furnace slag (GGBFS), natural pozzolan and limestone powder used to replace 6%, 20% and 35% (by mass) of PC with different finenesses.

PC particles are prone to show some coagulation when mixed with water. Incorporation of any ultra-fine mineral powder into the cementitious system causes a dispersion of the cement particles thus reducing the

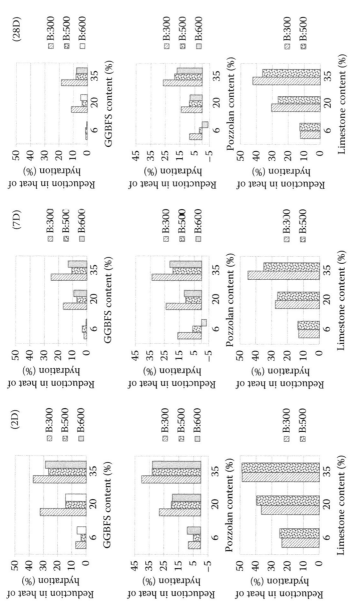

Figure 8.10 Relative reduction in heat of hydration at 2, 7 and 28 days when PCs of 300, 500 and 600 m²/kg Blaine specific surface are partially replaced by ground granulated BFS, natural pozzolan and limestone powder. (Adapted from Tokyay, M., Delibaş, T., Aslan, Ö. 2010. *Effects of mineral admixture type, Grinding Process, and Cement Fineness on the Physical and Mechanical Properties of GGBFS-, Natural Pozzolan-, and Limestone-incorporated Cements*, Working Paper AR-GE 2010/01-B, Turkish Cement Manufacturers' Association [TÇMB], 45p. [in Turkish].)

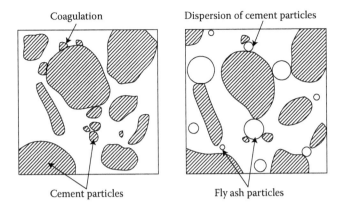

Figure 8.11 Dispersion effect of FA particles resulting in more cement particle surface exposed for hydration. More surface area of cement particles becomes available for reacting with water. (Adapted from Dhir, R.K. 1986. Pulverizedfuel ash, in *Cement Replacement Materials* [Ed. R.N. Swamy], pp. 197–256, Surrey University Press.)

tendency for flocculation and exposing more cement surface area for hydration (Dhir, 1986). At the same water–solids ratio, the water–PC ratio becomes higher and allows more space for the hydration of clinker phases (Lothenbach et al., 2011). Furthermore, the dispersion effect leads to a more homogeneous distribution of water within the cement paste and thus facilitates the hydration. In other words, for a fixed cement amount, mineral admixture incorporation would result in more cement hydration. This effect is illustrated in Figure 8.11 for FA-incorporated cements. Water availability for hydration is also achieved by the clogging of capillary channels in the fresh paste by fine mineral admixture particles. Reduced bleeding due to the prevention of water movement in the capillarities leads to more water being available for hydration. Another physical effect related with modification of particle size distribution by mineral admixtures is the possible reduction in the thickness of initial layer of hydrates formed on the surface of the cement particles. Such a reduction in the thickness of the inital hydrate layer makes breaking it down easier thus accelerating the hydration process. The phenomenon is illustrated in Figure 8.12.

The presence of mineral admixtures facilitates the early hydration of PC by providing additional sites for the precipitation of hydration products (Halse et al., 1984; Dhir, 1986; Lawrence et al., 2003). It is related with the fineness of the mineral admixture and its affinity for cement hydrates.

All four physical effects discussed earlier are true for any kind of mineral admixture whether it is non-reactive, pozzolanic or hydraulic. Mineral admixtures enhance the hydration of the PC portion of the cementitious system in which they are incorporated. However, this may be suppressed if the amount of mineral admixture used is high. Figure 8.13 compares the

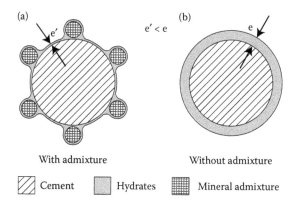

Figure 8.12 (a) and (b) Schematic representation of having thinner initial hydrate layer by mineral admixture incorporation. (Adapted from Lawrence, P., Cyr, M., Ringot, E. 2003. *Cement and Concrete Research*, 33(12), 1939–1947.)

24 h heat of hydration curves of natural pozzolan-incorporated cements with the control PC of the same fineness. The figure indicates that small amount of pozzolan incorporation obviously results in higher heat evolution and larger amounts lower the heat evolved. However, the reduction is less than that obtained from the remaining PC if the pozzolan were not effective at the early stages of hydration, as illustrated in Figure 8.14. In two investigations on the effects of FA on hydration, similar comments were made: It was found that after 28 days of hydration, the four cements

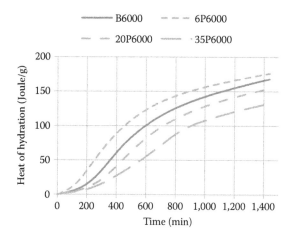

Figure 8.13 24 h heat of hydration curves of 6%, 20% and 35% natural pozzolan incorporated cements compared with their control PCs of the same fineness. (Data from Över, D. 2012. Early heat evolution in natural pozzolan-incorporated cement hydration, MS thesis, Middle East Technical University, 83pp.)

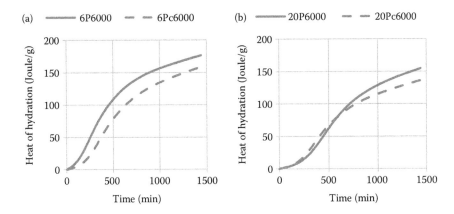

Figure 8.14 24 h measured heat of hydration curves of (a) 6% and (b) 20% natural pozzolan incorporated cements (full lines) compared with heat of hydration curves calculated with the assumption that there is no effect of mineral admixture on hydration (dashed lines). (Data from Över, D. 2012. Early heat evolution in natural pozzolan-incorporated cement hydration, MS thesis, Middle East Technical University, 83pp.)

containing 25% FA resulted in 3%–9% more alite (C_3S) hydration than the PC alone (Abdulmaula and Odler, 1981; Dalziel and Gutteridge, 1986).

8.3.2 Chemical influence

Among the three major groups of mineral admixtures, pozzolans and latent hydraulic materials are also involved in the hydration process chemically. This may be observed even for some mineral admixtures of low reactivity. Pozzolanic and/or hydration reactions of the mineral admixtures and their interactions with the hydration of PC form the three main parts of their chemical involvement in the hydration process. The rate and extent of this chemical involvement depends on their (1) chemical and (2) mineralogical compositions, (3) amount of glassy phase, (4) fineness, (5) amount, (6) the characteristics of the cement that they are used with and (7) ambient temperature and humidity.

Although the products obtained on pozzolanic and/or hydration reactions of chemically reactive mineral admixtures are similar in nature, due to their differences in type and composition, they are individually dealt with in this chapter.

8.3.2.1 Pozzolanic reactions

The general term pozzolanic activity refers to all chemical reactions between pozzolans, lime and water. The variability in the types, chemical and mineralogical compositions of the pozzolans makes it very difficult, if not impossible, to model the pozzolanic activity. Therefore, generally two parameters are used to describe the activity of any pozzolan: (1) lime-combining capacity and (2) rate of lime combination (Massazza, 1988).

Lime-combining capacity of a pozzolan depends on the following:

1. The nature and amount of the active phases present
2. The silica content
3. Relative proportions of lime and pozzolan in the mix
4. Curing period

On the other hand, the rate of lime combination by a pozzolan depends on

1. Fineness of the pozzolan
2. Water–solids ratio of the pozzolan–lime–water mixture
3. Ambient temperature

8.3.2.1.1 Natural pozzolans

Many different minerals may be present in natural pozzolans. However, the fundamental constituent is the glassy phase of siliceous nature. The average amount of glass in a pyroclastic pozzolan is usually greater than 50%. It may be as low as 25% in clastic rocks. The rest are different clay minerals, quartz and feldspars. Lime-combining capabilities of different phases of a natural pozzolan (Rhenish trass) as reported by Ludwig and Schwiete (1963) are given in Table 8.4. It is obvious that the glassy phase combines the majority of the lime. However, in that same study glasses in three different pozzolans were found to have different lime-combining capabilities.

It is evident that the lime-combining capacity of a pozzolan depends on the relative proportions of the mineral phases present. Since glass is the dominant phase, increase in its amount leads to higher pozzolanic activity.

Although the mineral composition of a natural pozzolan is the most significant parameter for lime combination, the amount of lime combined

Table 8.4 Lime-combining capabilities of different mineral phases in a natural pozzolan

Mineral phase	Lime-combining capacity (mgCaO/g)	Amount in the pozzolan (%)	Calculated lime combination in the pozzolan (mgCaO/g)
Glass	364	55	200.0
Quartz	43	13	5.5
Feldspar	117	15	17.5
Leucite	90	6	5.4
Analcime	190	7	13.3
Kaolinite	34	2	0.7
Total	–	98	242.5

Source: Adapted from Ludwig, U., Schwiete, H.E. 1963. Zement-Kalk-Gips, 10, 421–431.

by a pozzolan can be also related to its silica (SiO_2) content. Higher silica content generally indicates higher capacity of lime combination. However, consideration of only the silica content may be misleading from the pozzolanic activity point of view since there are cases of natural pozzolans of higher SiO_2 resulting in less lime combination than those with lower SiO_2 (Massazza, 1988).

The amount of lime combined by pozzolan increases the lime–pozzolan ratio of the mix and the curing time increases (Takemoto and Uchikawa, 1980; Massazza, 1988).

The rate at which the lime is combined by a pozzolan changes with age and is dependent more on the fineness of the pozzolan at early reaction stages (Takemoto and Uchikawa, 1980). In an experimental study by Shi and Day (2000), various different methods for enhancing the reactivity of a natural pozzolan were investigated by using a 20:80 (by mass) mixture of $Ca(OH)_2$ and a volcanic ash. The effect of pozzolan fineness on lime combination was studied for three different Blaine specific surface areas of pozzolan as 291, 385 and 554 m^2/kg. Their results, as shown in Figure 8.15, support the previous statement by Takemoto and Uchikawa (1980).

The pozzolanic reaction is accelerated as the water content of the pozzolan–lime mix increases and as with all chemical reactions, the rate of lime combination by pozzolans increases with increasing temperature. However,

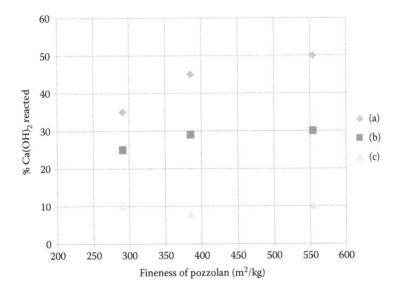

Figure 8.15 Amount of $Ca(OH)_2$ reacted in 20:80 lime–pozzolan mixtures at different stages of reaction: (a) within first day, (b) between 1 and 3 days and (c) between 3 and 7 days. (Data from Shi, C., Day, R.L. 2000. *Cement and Concrete Research*, 30(4), 607–613.)

this beneficial effect of increased temperature seems to stop or decrease at temperatures above 60°C (Collepardi et al., 1976).

Pozzolanic reaction products are similar to the hydration products of PC. Natural pozzolans react with lime, under moist conditions, and result in calcium silicate and calcium aluminate hydrates. Besides C–S–H and C_4AH_{13}, gehlenite hydrate, hydrogarnet and calcium carboaluminate hydrate had been observed in pozzolan–lime water mixtures (Sersale and Rebuffat, 1970; Takemoto and Uchikawa, 1980). Generally, the latter appear much later than the former two. Common products that form on pozzolanic reaction are listed in Table 8.5.

Pozzolanic materials have much less calcium content than PCs. Pozzolanic reactions in portland–pozzolan blends first require the formation of CH on the hydration of calcium silicates in the PC. Once the CH formation starts, siliceous and aluminous components in the pozzolan begin to react with this hydration product. Consequently, the CH content of hydrated pozzolan-incorporated cements is always lower than that of hydrated PCs. Besides the pozzolanic reactions, the presence of pozzolans also modifies the hydration kinetics and the hydration products of PC. Much of the recent works on the pozzolanic and/or latent hydraulic activities of mineral admixtures are concerned with PC–FA, PC–SF and PC–BFS reactions. These will be discussed in the relevant parts of this chapter.

Individual reactions of the natural pozzolan with the major compounds of PC would be a good starting point since the hydration of PC by itself is complex and pozzolan incorporation makes it even more complex and difficult to follow and quantify.

Table 8.5 Reaction products in pozzolan–lime-water mixtures

Hydrate	Explanation
C–S–H	Calcium silicate hydrate gel appears within first few days and increases as the reactions continue. Its C–S ratio may range 0.75–1.75, depending on the type of pozzolan, curing period and temperature and lime–pozzolan ratio.
C_2ASH_8	Gehlenite hydrate (stratlingite) is usually observed beyond 90 days of curing.
C_4AH_{13}	Small quantities of tetracalcium aluminate hydrate may be observed as early as calcium silicate hydrate.
C_3AH_6	C_4AH_{13} may convert into hydrogarnet at high temperatures or it may be observed after the combined lime is about 50% of the initial mass of the pozzolan.
$C_4A\bar{C}H_{12}$	Calcium carboaluminate hydrate may result from the reaction of carbonated lime and the aluminuous phase in the pozzolan.

Source: Adapted from Takemoto, K., Uchikawa, H. 1980. Hydration of pozzolanic cements, *Proceedings 7th International Congress on Chemistry of Cement*, Vol. I, Sub-theme IV-2, pp. 1–21, Paris; Massazza, F. 1988. Pozzolana and pozzolanic cements. in *Lea's Chemistry of Cement and Concrete*, 4th Ed. (Ed. P.C. Hewlett), Elsevier, Oxford; Sersale, R., Rebuffat, P. 1970. *Zement-Kalk-Gips*, 23(4), 182–184.

Ogawa et al. (1980) proposed that the hydration of C_3S is accelerated by natural pozzolan. They found out that the dormant period does not change significantly and the C_3S hydration peak is slightly delayed with respect to that of the control C_3S but its intensity is increased considerably.

While the acceleration of the early hydration of C_3S can be attributed to the physical influence of fine pozzolan particles, it increases, although at a much slower rate, with age. Later age C_3S hydration depends largely on the type of pozzolan used. However, differences become negligible beyond 90 days as shown in Figure 8.16.

On contact of C_3S–pozzolan blends with water, there results a very rapid formation of C–S–H both on C_3S and pozzolan particles. The C–S–H formed on pozzolan particles is thinner, more porous and has a much lower lime–silica ratio (C/S) than that formed on C_3S particles. After about 3 days of hydration, C/S close to the pozzolan particles significantly increases, indicating the presence of CH. H_2O in the basic pore solution facilitates the dissolution of sodium and potassium ions from the pozzolan particles and a silica and alumina-rich layer forms on their surfaces. Calcium ions present in the pore solution react with this layer to form calcium silicate and calcium aluminate hydrates. A mechanism for C_3S–pozzolan reaction proposed by Uchikawa and Uchida (1980) is given schematically in Figure 8.17.

Sufficient data regarding the effect of natural pozzolans on the hydration of C_2S–pozzolan system are not available.

The interaction of pozzolans with C_3A hydration was studied by various researchers in systems that include gypsum and/or CH (Collepardi et al., 1976;

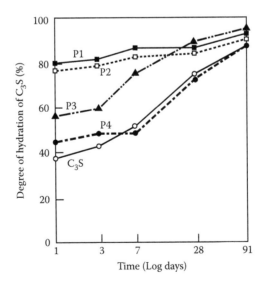

Figure 8.16 Degree of hydration of C_3S in 6:4 blends of C_3S–natural pozzolan for four different pozzolans. (Adapted from Ogawa, K. et al. 1980. *Cement and Concrete Research*, 10(5), 683–696.)

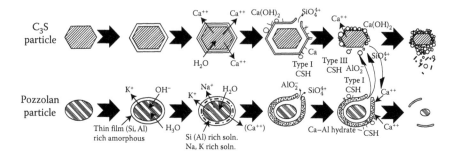

Figure 8.17 Schematic representation of the development of C₃S–pozzolan reaction. (Adapted from Ogawa, K. et al. 1980. *Cement and Concrete Research*, 10(5), 683–696.)

Holten and Stein, 1977; Uchikawa and Uchida, 1980). Calcium aluminum hydrates that are formed in C₃A–pozzolan mixes are the same as those that form upon the hydration of C₃A alone (Massazza, 1988). A similar statement holds true for C₄AF–pozzolan mixtures. On the other hand, if the mixture contains CH, additional C–S–H forms as a result of pozzolanic reaction. A mechanism for C₃A–pozzolan interaction with the existence of gypsum and CH, proposed by Uchikawa and Uchida (1980) is given schematically in Figure 8.18.

8.3.2.1.2 Low-lime fly ashes

Low-lime fly ashes consist mainly of silica (SiO_2). However, some may contain a significant amount of alumina (Al_2O_3). The amount of lime (CaO) is limited. The glassy phase is usually over 50% of the total mass and may be as high as 95%. Although the pozzolanic reaction products of low-lime

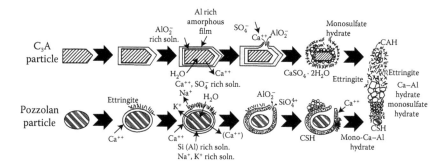

Figure 8.18 Schematic representation of the development of C₃A–pozzolan–gypsum–CH interaction. (Adapted from Uchikawa, H., Uchida, S. 1980. *Proceedings of the 7th International Congress on Chemistry of Cement*, Vol. 3, pp. IV/24–IV/29, Paris.)

fly ashes are essentially the same as those obtained from natural pozzolans, the presence of a considerable amount of alumina leads to Al-rich phases (Collepardi et al., 1978). Furthermore, if the sulphate ion (SO^{4-}) concentration of the FA is high, it may react with the aluminous components in the FA in lime water to form ettringite or monosulphate form, depending on the sulphate–alumina ratio (Snellings et al., 2012). Gypsum may also precipitate from the lime water (Massazza, 1988). Low-lime fly ashes, when mixed with lime water form C–S–H, C_4AH_{13}, C_2SAH_8 and sometimes calcium carboaluminates (Takemoto and Uchikawa, 1980; Aïtcin et al., 1986). Pozzolanic reaction products in the $CaO–SiO_2–Al_2O_3$ system are shown in the ternary diagram given in Figure 8.19.

The blending of PC with low-lime fly ash results in lower CH in the hydrated paste due to (1) reduced amount of calcium silicates in the PC by partial replacement with FA and (2) its consumption by pozzolanic reactions to produce more C–S–H, at later ages.

The C–S–H phase in low-lime fly ash incorporated cements has lower Ca–Si ratios (1.0–1.8) than those of PCs (1.2–2.1) due to higher availability of silica obtained by the dissolution of the FA (Richardson, 1999; Snellings et al., 2012). Furthermore, fly ashes usually have high amounts (10%–35%)

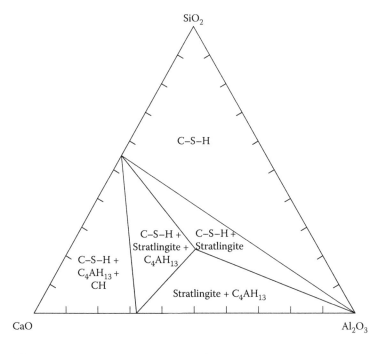

Figure 8.19 Pozzolanic reaction products of $CaO–SiO_2–Al_2O_3$ ternary systems. (Adapted from Snellings, R., Mertens, G., Elsen, J. 2012. *Reviews in Mineralogy & Geochemistry,* 74, 211–278.)

of alumina, so using such fly ashes leads to alumina-rich hydrated phases. Hydration of PC-low lime fly ash blends was modelled with the assumption that Al–Si ratio in C–S–H is 0.10 (Lothenbach et al., 2011). On full hydration of the PC portion and 50% reaction of the FA portion, CH (portlandite) was calculated to be consumed to form additional C–S–H with lower lime–silica ratio, that is, the formation of tobermorite-like C–S–H besides the jennite-like C–S–H which has higher lime–silica ratio. Calcium carboaluminate hydrate and ettringite get unstable and stratlingite (calcium–aluminum–silicate hydrate) forms if the amount of FA is more than 30%. For stratlingite to form, FA should be Al-rich. The hydration and pozzolanic reaction products of PC–low-lime fly ash blends based on the model mentioned earlier are shown in Figure 8.20.

Most of the outcomes of the model proposed by Lothenbach et al. (2011) agree with the experimental results of other researchers (Hanehara et al., 2001; Dyer and Dhir, 2004; Escalante-Garcia and Sharp, 2004; Luke and Lachowski, 2008).

The rate of pozzolanic reaction of low-lime fly ash is slow at ordinary temperatures. A noticeable decrease in the amount of portlandite in hydrated PC–FA blends can be observed after at least a week (Hanehara et al., 2001; Lothenbach et al., 2011). Slightly higher amounts of portlandite may be observed in the earlier stages of hydration due to the filler effect of FA that accelerates the C_3S hydration (Lam et al., 2000; Hanehara et al., 2001; Pane and Hansen, 2005; Sakai et al., 2005).

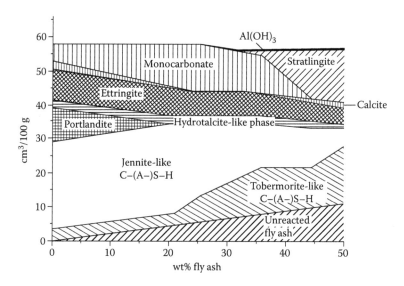

Figure 8.20 Modelled hydration and pozzolanic reaction products of PC-low-lime fly ash blends. (From *Cement and Concrete Research*, 41(3), Lothenbach, B., Scrivener, K., Hooton, R.D. Supplementary cementitious materials, 311–323, Copyright 2011, with permission from Elsevier.)

8.3.2.1.3 Silica fume

SF contains more than 80% silica in almost completely amorphous state and it is in the form of extremely fine spherical particles (~0.10 μm average diameter). Its high silica content, extreme fineness and high degree of amorphousness make it a very reactive pozzolan. The pozzolanic reaction products of SF are essentially the same as those of natural pozzolans and low-lime fly ashes. However, they form at a much faster rate. In a comparative study that investigated pozzolanic activities of a SF, a low-lime fly ash, a natural pozzolan and a zeolite (clinoptilolite) 1:1 CH–pozzolan pastes were prepared with 0.55 water–solids ratio and CH depletion was monitored at 3, 7 and 28 days through thermogravimetric analysis (TGA) (Uzal et al., 2010). The results shown in Figure 8.21 clearly indicate that both the lime-combining capacity and rate of lime combination of SF are higher than those of the other pozzolans. On the other hand, it is necessary to note that Brunauer–Emmett–Teller (BET) specific surface areas of the materials used were 35,500, 24,650, 18,780 and 3,380 m²/kg for zeolite, natural pozzolan, SF and FA, respectively (Uzal et al., 2010).

The reactivity of SF in PC–SF blends is also high. However, pozzolanic reactions may continue for months (Hjorth et al., 1988; Pietersen et al., 1992; Pane and Hansen, 2005). The reaction rate of SF depends on the alkalinity of the pore solution. As the pH of the pore solution strongly increases within 12–16 h of hydration the rate of reaction of SF gets accelerated and upto 80% of all SF incorporated may react within the first 3

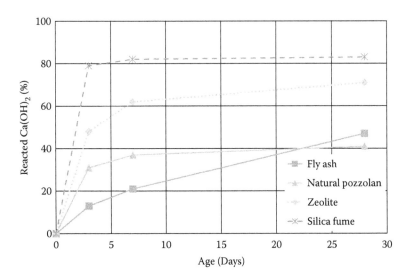

Figure 8.21 Lime-combining rate and capacity of SF in comparison with other pozzolans. (Data from Uzal, B. et al. 2010. *Cement and Concrete Research*, 40(3), 398–404.)

days (Barneyback and Diamond, 1981; Hjorth et al., 1988; Lothenbach et al., 2011). Rapid pozzolanic reaction of SF results in lowered pH of the pore solution upon which the rate of reaction is slowed down (Shehata and Thomas, 2002; Lothenbach et al., 2011).

8.3.2.2 Latent hydraulic reactions

GGBFS and high-lime fly ashes can behave as hydraulic binders if they have an appropriate chemical composition. However, these materials usually need suitable activators to exhibit hydraulic reactivity, unlike PCs which alone react readily with water. This is why they are called latent hydraulic materials.

8.3.2.2.1 Ground granulated blast furnace slag

When GGBFS is mixed with water, a thin layer of silica-rich calcium silicate hydrate with low lime content forms on the surface of the particles. This hydrate layer prevents further contact of water with slag and the ionic dissolutions from the slag. Thus, the hydration stops (Regourd et al., 1983; Odler, 2000). For a continuing hydration, either alkaline or sulphatic activators should be added to the GGBFS–water mixture. Alkaline activators such as $Ca(OH)_2$, NaOH, KOH and Na_2SiO_3 increase the alkalinity of the liquid phase of the mixture (pH > 12), thus preventing the formation of the impermeable layer and facilitating the dissolution of ions to result in the precipitation of hydration products. On the other hand, calcium sulphates (gypsum, hemihydrate, anhydrite) attract Ca^{2+} and $Al(OH)^{4-}$ ions released from the slag into the pore solution and result in ettringite formation besides C–S–H (Odler, 2000).

PC–GGBFS blends contain both alkaline and sulphatic activators. CH from the hydration of C_3S, sodium and potassium hydroxides that form from the alkalies in the PC portion and gypsum which is present in the PC as a retarder are the activators for the hydration reactions of GGBFS. Lothenbach et al. (2011) thermodynamically modelled the hydration of PC–GGBFS blends and calculated the amount of hydration products as shown in Figure 8.22. The proposed model agrees with the experimental research. On full hydration of the PC portion and 75% reaction of the slag portion, most of the CH (portlandite) was calculated to be consumed to form additional C–S–H with lower lime–silica and higher alumina–silica ratios than that formed by PC hydration, alone. At sufficiently high GGBFS contents, portlandite may be completely consumed. However, at the early stages of hydration, the amount portlandite formed may be slightly more than that formed by PC hydration alone, due to the filler effect of GGBFS that accelerates C_3S hydration (Escalante-Garcia et al., 2001; Pane and Hansen, 2005). Although GGBFS contains a considerable amount of Al_2O_3,

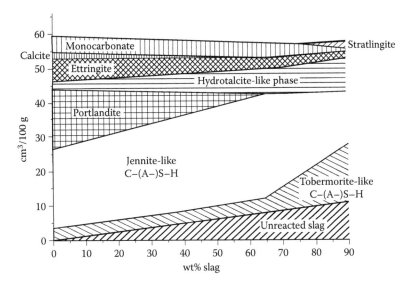

Figure 8.22 Modelled hydration and pozzolanic reaction products of PC–GGBFS blends. (From *Cement and Concrete Research*, 41(3), Lothenbach, B., Scrivener, K., Hooton, R.D. Supplementary cementitious materials, 311–323, Copyright 2011, with permission from Elsevier.)

ettringite (Aft) and monosulphate form (Afm) are usually less than that that would form on PC hydration (Richardson, 1999; Luke and Lachowski, 2008). The reduced amount of calcium aluminosulphates is attributed to the high alumina uptake of the C–S–H gel formed (Richardson and Groves, 1997; Taylor et al., 2010). The relatively high MgO contents in GGBFS results in more a hydrotalcite-like phase (Lothenbach et al., 2011).

8.3.2.2.2 High-lime fly ash

High-lime fly ashes of suitable composition may have self-cementitious properties (Joshi and Ward, 1980; Liu et al., 1980; Sullentrop and Baldwin, 1983; Papayianni, 1992; Tokyay and Hubbard, 1992). Their hydration depends on the solubility of both the amorphous and crystalline portions of the ash as they determine the concentration of the ions released into the pore solution and thus affect the types of hydration products and cementitious properties (Tishmack et al., 2001). The mineral phases in the fly ashes are mainly governed by the types of impurities in the coals from which the fly ashes are obtained (Hubbard and Dhir, 1984). The low-lime fly ashes that are usually obtained from bituminous coals are composed of crystal phases of quartz, mullite, hematite and magnetite in the aluminosilicate glassy matrix (Gomes and François, 2000). On the other hand, high-lime fly ashes which are obtained from lignites or subbituminous coals may contain

Ca-bearing crystalline compounds such as free lime, anhydrite, plagioclase, gehlenite, etc. Furthermore, the amorphous or glassy phase of these fly ashes may be calcium alumino silicate or even calcium aluminate type rather than aluminosilicate type that is generally observed in the low-lime fly ashes.

In an experimental investigation (Tokyay and Hubbard, 1992) on the self-cementitious values of three high-lime fly ashes, the relationships found between the impurities in the coals and mineral phases in their respective fly ashes are summarised in Table 8.6. Quartz and hematite in the coals withstand the combustion heat and directly pass to the FA, keeping their original structure. The presence of anhydrite in the fly ashes is due to two different reasons: (1) Coals 1 and 3 contain aragonite and calcite. Aragonite transforms into calcite at around 400–500°C during burning and at 900°C calcite decomposes into free lime and CO_2. On the other hand, pyrite in the coals transforms into magnetite upon oxidation and gives off SO_2, which then combines with some of the free lime to produce anhydrite. (2) Coal 2 contains gypsum as an impurity and it dehydrates to result in anhydrite in the FA.

Generally, high-lime fly ashes contain low proportions of glassy component when compared with the dominating values in the low-lime fly ashes. The other component of the amorphous phase is the residual clay. For combustion temperatures around 1000–1100°C in thermal power plants, the clay minerals lose their crystallinity and transform into reactive silica and alumina (Tokyay and Hubbard, 1992).

Depending on the free lime and anhydrite contents of the high-lime fly ashes, the first hydration product that precipitates may be ettringite. Free lime transforms into CH and anhydrite transforms into gypsum in the

Table 8.6 Mineralogical compositions of three high-lime fly ashes in relation with the impurities in the coals

Impurities in the coals			Mineralogical composition of fly ashes (%)			
Coal 1	Coal 2	Coal 3	Mineral	HLFA 1	HLFA 2	HLFA 3
Aragonite	Siderite	Calcite	Quartz	4.5	5.6	5.1
Calcite	Fluoroapatite	Aragonite	Lime	18.6	5.5	9.8
Montmorillonite	Chlorite	Illite	Anhydrite	12.2	9.3	7.4
Hematite	Illite	Chlorite	Plagioclase[a]	~28	~15	~20
Quartz	Gypsum	Quartz	Hematite	4.0	6.0	2.0
Pyrite	Quartz	Pyrite	Magnetite	0.8	2.5	0.6
	Pyrite		Mullite	1.0	1.2	4.3
	Hematite		Amorphous	~30	~50	~50

Source: Adapted from Tokyay, M., Hubbard, F.H. 1992. Mineralogical investigations of high-lime fly ashes, 4th CANMET/ACI International Conference on FlyAsh, Silica Fume, Slag and Natural Pozzolans in Concrete (Ed. V.M. Malhotra), Vol. 1, pp. 65–78, Istanbul.

[a] Includes gehlenite, also.

early stages of hydration. At the start of these transformations, ettringite begins to form due to the reaction between the amorphous alumina, gypsum and CH. The transformations and reactions mentioned earlier may continue upto 7 days. Hydration products of HLFA 1 and HLFA 3 of Table 8.6 were examined by XRD and no gypsum and CH peaks were observed beyond 7 days of hydration indicating a complete consumption of these to form mainly ettringite. Besides ettringite, C–S–H and calcite formations were also observed as early as 3 days of hydration. At 90 days, $C_4A\overline{C}H_{11}$ and C_4AH_{13} were observed together with the aforementioned products. Although HLFA 2 is also a high-lime fly ash, it differs from the other two in the sense that it contains much less free lime and the only hydration product that is formed when it is mixed with water was gypsum. However, upon additional CH incorporation, it behaved in a similar manner as the other two high-lime fly ashes (Tokyay and Hubbard, 1992).

Chapter 9

Effects of mineral admixtures on the workability of fresh concrete

9.1 GENERAL ABOUT CONCRETE WORKABILITY

The properties of fresh concrete are important since they affect the properties of hardened concrete. For a concrete to be of a desired quality in hardened state for a specific set of job conditions, it must fulfil the following requirements in fresh state (Mindess and Young, 1981):

1. Easily mixed and transported
2. Be uniform both within batch and between batches
3. Have flow properties that makes it fill the forms properly
4. Be compacted sufficiently without excessive effort
5. Not be segregated during placing and compaction
6. Be finished properly

It is obvious that all these requirements are subjective and qualitative. Therefore, the term *workability* which embodies all the properties mentioned earlier is preferred. Workability of concrete is defined as that property determining the effort required to place, compact and finish a freshly mixed quantity of concrete with a minimum loss of homogeneity (ASTM C 125, 2000). Workability consists basically of *consistency* and *cohesiveness* of the fresh concrete. In other words, fresh concrete must be mobile and yet sufficiently stable. Mobility of fresh concrete which is determined by measuring its consistency depends largely on its 'wetness' whereas its stability has two aspects: water-holding and coarse aggregate-holding abilities. The former shows the resistance to *bleeding* and the latter shows the resistance to *segregation*. Both phenomena result in the separation of concrete components in fresh state during placing and compaction. Segregation is the separation of coarse aggregate particles from the concrete body and/or settlement of heavy particles whereas bleeding is the appearance of some of the water in fresh concrete on the surface after compaction and finishing.

The workability of concrete is affected by

1. Water content of the mix
2. Relative proportions of concrete ingredients
3. Aggregate properties such as shape, texture, porosity
4. Time and temperature
5. Characteristics of the cement and admixtures used

A concrete mix that cannot be placed and compacted properly is not likely to show the desired mechanical and durability characteristics at later ages. On the contrary, fresh concrete that may be workable for a certain job may not be workable for another one. Therefore, workability must always be related to the type of construction and method of placement, compaction and finishing (Mehta and Monteiro, 2006).

9.1.1 Basic principles of rheology related with fresh concrete

All simple liquids obey Newton's law of viscous flow which can be explained by referring to the laminar flow illustrated in Figure 9.1.

On applying a tangential force, F on plane A, it moves with respect to the stationary plane B and carries the parallel layers of liquid between A and B. The velocity of the liquid particles in each layer is a function of the distance, l between the planes A and B. Newton expressed the relationship between the tangential force and the velocity gradient, dV/dl by the equation

$$F = \eta \frac{dV}{dl} A \quad \text{or} \quad \tau = \eta \frac{dV}{dl} \tag{9.1}$$

where η is the coefficient of viscosity.

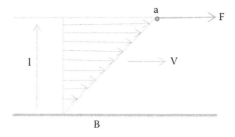

Figure 9.1 Laminar flow to describe Newton's law of viscous flow.

The flow rate can also be expressed by the time rate of shear strain,

$$\frac{d\gamma}{dt}\tau = \eta\frac{d\gamma}{dt} \tag{9.2}$$

When the time behaviour of flow is concerned, rate of strain is considered. On the other hand, when the geometric aspect of flow is concerned, velocity gradient is used (Jastrzebski, 1959). The liquids that show this behaviour are called Newtonian liquids. The stress–rate of strain relationship of Newtonian liquids is linear as illustrated in Figure 9.2.

Newton's law of viscous flow applies to dilute suspensions of solid particles in simple liquids, as well. The only effect of suspended solid particles is that the coefficient of viscosity increases as the amount of suspended particles increase. The flow behaviour of highly concentrated suspensions is different from the Newtonian flow. Bingham showed that the flow of such materials can be represented by the equation

$$\tau - \tau_y = \eta\frac{d\gamma}{dt} \tag{9.3}$$

where τ_y is the yield stress. Such behaviour is shown in Figure 9.3. Flow takes place only at stresses above the yield value. It also follows that upto the yield stress, the material behaves as an elastic solid which is able to carry loads without permanent deformation. Materials that show this behaviour are named Bingham bodies (Jastrzebski, 1959).

The flow behaviour of fresh concrete which is a heavily concentrated suspension can roughly be approximated by the Bingham model. However, the actual rheology of cementitious systems may be very complex. Shear-thinning or shear-thickening phenomena are usually observed and the rheological properties are time dependent. An important aspect of flow of fresh concrete that may be significant when placing and compacting is its thixotropic character (Mindess and Young, 1981). Thixotropy is the property

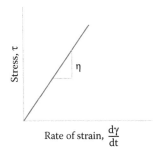

Figure 9.2 Stress–rate of strain relationship of Newtonian liquids.

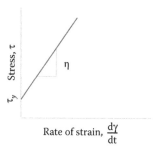

Figure 9.3 Flow behaviour of the Bingham body.

of certain suspensions which results in a decrease in coefficient of viscosity under stress. When the stress is removed, the resistance to flow is gradually recovered (Jastrzebski, 1959). Sometimes, anti-thixotropy that is the increase in viscosity under stress may also be encountered. Furthermore, there are numerous parameters of cementitious materials such as (1) interparticle forces like Brownian, van der Waals, viscous and liquid bridging forces, (2) particle shape, (3) particle size distribution and fineness and (4) surface charge which may result in the flocculation or dispersion of particles which in turn, affect the rheological behaviour of the cementitious systems (Nehdi et al., 1998; Moosberg-Bustnes, 2003). Besides all these, test methods and conditions may also complicate the situation and sometimes lead to misleading conclusions.

9.1.2 Various methods for measuring concrete workability

There are numerous different test methods proposed to measure the workability of fresh concrete. Only a few of those were standardised and almost all of them are empirical. Nevertheless, it is significant to have some measure of workability and the available measurement methods, at least, provide information on the variations of this property. It must be known that none of the methods can be used to measure the workability of the whole range of fresh concretes from very stiff to very fluid.

The methods to measure the workability of fresh concrete are grouped into different categories as (a) slump test, (b) compaction tests, (c) flow tests, (d) remolding tests, (e) penetration tests, (f) mixer tests (Mindess and Young, 1981) and (g) concrete rheometres (Banfill et al., 2001).

The *slump test* is the most popular method due to its simplicity. However, it is basically a consistency measurement and may not be suitable for very dry or very wet concretes. It may be considered as the measurement of the resistance of concrete to flow under its own weight. The *compaction tests* measure the compactibility of fresh concrete for a specified amount of work. The most common method in this category is the *compacting factor test* which measures the degree of compaction by relating the density of uncompacted fresh concrete to that of the compacted one. *Flow tests* measure the flowability of fresh concrete on jolting or vibration. They may provide some information about the segregation tendency, also. *Remolding tests* were developed to measure the work required to make the concrete flow and change its shape from a frustum of a cone to a cylinder. *VeBe test* and *Thaulow Drop Table test* are the common examples for this category. The time required for the shape change stated previously on applying a vibration of controlled frequency and amplitude is recorded in the former, whereas the number of drops to achieve the shape change is recorded in the latter. *Penetration tests* measure the depth of penetration of an indenter into concrete. The most commonly used method in this category is the *Kelly Ball test* which measures the penetration of a hemispherically shaped apparatus weighing 13.6 kg into concrete. *Mixer tests* measure the power required to turn a concrete mixer filled with concrete. Although they may give more information on the rheological characteristics, tests in this category are not practical in site conditions (Mindess and Young, 1981; Mehta and Monteiro, 2006). As the demand for control of workability increased with the development of new and special concretes such as self-compacting concrete (SCC) and high-performance concrete, more sophisticated tools were started to be designed. The devices which are called as *concrete rheometers* make use of rheology methods to measure the flow properties of fresh concretes. Basically, they measure the shear stress at varying shear rates (Banfill et al., 2001).

9.2 CHANGES IN CONCRETE WORKABILITY UPON MINERAL ADMIXTURE INCORPORATION

Mineral admixtures typically have lower densities than PCs. Therefore, whether they are used as a partial cement replacement material or addition in concrete, the amount of paste increases for given water content and the plasticity and cohesiveness of the fresh concrete consequently increases. On the other hand, there are many other characteristics of individual mineral admixtures that would affect the workability of mineral admixture-incorporated concretes. Therefore the extent of the changes in workability through mineral admixture use should be considered separately for different types of these materials.

9.2.1 Water demand

Effect of mineral admixtures on workability of concrete is commonly related with the changes they cause in the water demand of the mixture. Generally, for a constant volume of cementitious material, mineral admixture-incorporated concretes show less workability than similar PC concretes. The most common reason for this is the increase in the water demand that the mineral admixtures cause due to their higher specific surface area (Ferraris et al., 2001). The increase in the surface area corresponds to an increase in the interparticle surface forces that result in the higher cohesiveness of the mix which also means lower mobility. Thus, for a specified workability, usually water content must be increased when fine mineral admixtures are used. This effect is always observed in concretes containing SF and this is the reason why SF should be used together with plasticising or superplasticising chemical admixtures (Fidjestøl and Lewis, 1988). However, there are many cases in the literature that report reduced water demand on mineral admixture use. Most of such reports are on low-lime fly ashes and some on SF. The hypotheses related with the improved workability are based on the spherical particle shapes of these materials: (1) the spherical particles roll over each other and reduce the interparticle friction (Ramachandran, 1995) and act as ball bearings giving the mix more mobility (Fidjestøl and Lewis, 1988); (2) having the shape that has the least surface area–volume ratio, the particles result in less water requirement for wetting; and (3) spherical particles result in denser packing than angular particles, in wet state which causes lower water retention therefore, lower water demand (Sakai et al., 1997). The idea of improved particle packing was also used in several reports to explain the reduced water demand on GGBFS and limestone powder incorporations (Meusel and Rose, 1983; Ellerbrock et al., 1985; Lange et al., 1997).

9.2.1.1 Effect of natural pozzolans

Natural pozzolans have long been known to increase the water demand of pastes, mortars or concretes for a specified consistency (Davis et al., 1935). The water requirement increases with increasing level of PC replacement and with increasing fineness (Cook, 1986a). Examples for increased water demand with increasing natural pozzolan content are shown in Figures 9.4 and 9.5, which relate the pozzolan content of the pastes to the water requirement for normal consistency. The former is for a volcanic tuff and the latter is for a diatomaceous earth.

The combined effect of pozzolan content and cement fineness of interground and separately ground blended cements was investigated as part of a more comprehensive research on the characteristics of different mineral admixture-incorporated cements (Tokyay et al., 2010). The natural pozzolan used was a volcanic tuff that is being used commercially in blended cement manufacturing. The results obtained are illustrated in Figure 9.6.

9.2.1.2 Effect of fly ashes

The favourable influence of fly ashes on the water demand of concrete for a specified workability, was established as early as 1937 (Davis et al., 1937).

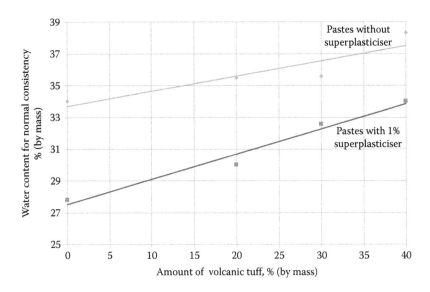

Figure 9.4 Change in water requirement for normal consistency with increased amount of volcanic tuff in the cement paste. (Data from Colak, A. 2003. *Cement and Concrete Research*, 33(4), pp. 585–593.)

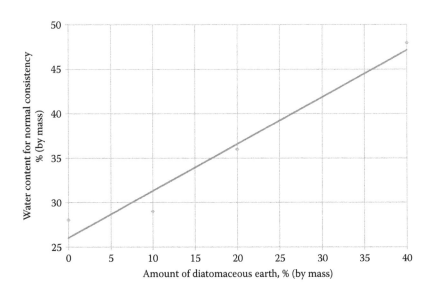

Figure 9.5 Change in water requirement for normal consistency with increased amount of diatomaceous earth in the cement paste. (Data from Aruntaş, H.Y. 1996. Usability of diatomites as pozzolans in cementitious systems, PhD thesis, Gazi University Institute of Natural and Applied Sciences [in Turkish].)

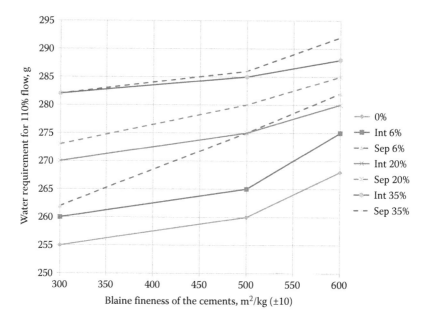

Figure 9.6 Change in water requirement of mortars prepared by using interground and separately ground blended cements with different natural pozzolan contents and Blaine-specific surface areas. (Adapted from Tokyay, M., Delibaş, T., Aslan, Ö. 2010. *Grinding Process, and Cement Fineness on the Physical and Mechanical Properties of GGBFS-, Natural Pozzolan-, and Limestone-incorporated Cements,* Working Paper AR-GE 2010/01-B, Turkish Cement Manufacturers' Association [TÇMB], 45p. [in Turkish].)

Since then, many investigators had verified this (Dhir, 1986; von Berg and Kukko, 1991). However, the water demand of a certain type of fly ash (FA), as well as its rheological effects in concrete, is dependent on many factors, the most significant of which are its amount, fineness, particle shape and loss on ignition (von Berg and Kukko, 1991). Examples on the effects of these factors are illustrated in Figures 9.7 through 9.10, respectively. It should be noted that different fly ashes were used in determining the effects of these individual factors which means that there had been other parameters involved. In fact, none of these factors are independent from each other.

9.2.1.3 Effect of GGBFS

Although GGBFS particles are angular in shape, they have hard and smooth surfaces. Thus, their much lower water absorption, if any, when compared with PC particles makes the water demand of GGBFS-incorporated concrete lower than that of PC concrete (Wainwright, 1986; ACI 233, 2003).

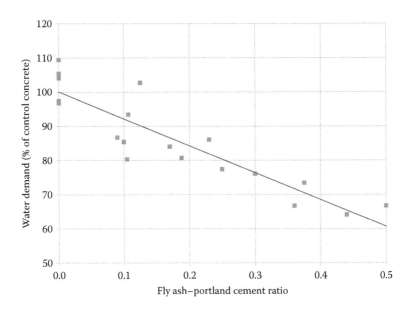

Figure 9.7 Water demand of the PC-FA blends in concrete. (Adapted from von Berg, W., Kukko, H. 1991. Fresh mortar and concrete with fly ash, in *Fly Ash in Concrete – Properties and Performance, Rept. Tech. Commun. 67-FAB RILEM* (Ed. K. Wesche), E. & F.N. Spon, London.)

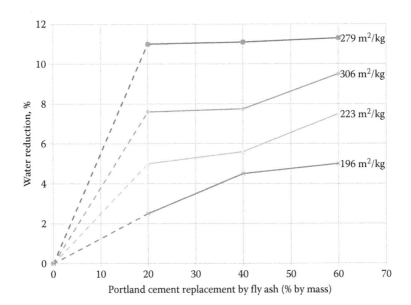

Figure 9.8 Water demand of PC-FA mortars for different Blaine fineness of fly ashes. (Adapted from Helmuth, R.A. 1986. *Mortars, and Concretes: Causes and Test Methods*, ACI SP-91, pp. 723–740.)

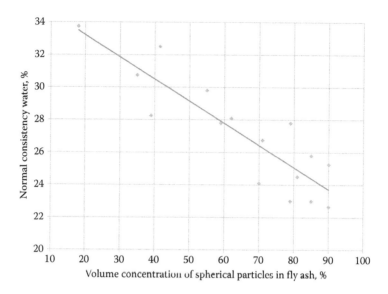

Figure 9.9 Change in normal consistency water requirement of 70:30 PC:FA pastes with increasing amount of spherical particles in the FA. (Adapted from Braun, H., Gebauer, J. 1983. *Zement-Kalk-Gips*, 36(5), 254–258.)

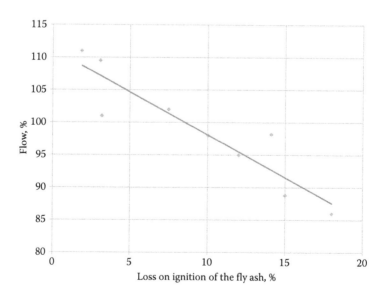

Figure 9.10 Effect of loss on ignition of FA on the flow of mortars made with blended cements having 25:75 FA:PC ratio, by mass. (Adapted from von Berg, W., Kukko, H. 1991. Fresh mortar and concrete with fly ash, in *Fly Ash in Concrete – Properties and Performance*, Rept. Tech. Comm. 67-FAB RILEM (Ed. K. Wesche), E. & F.N. Spon, London.)

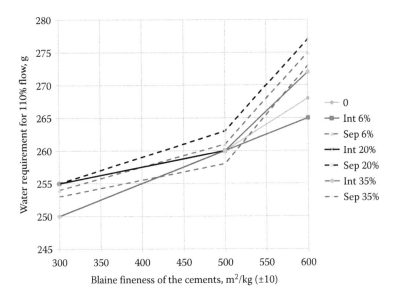

Figure 9.11 Change in water requirement of mortars prepared by using interground and separately ground blended cements with different GGBFS contents and Blaine-specific surface areas. (Adapted from Tokyay, M., Delibaş, T., Aslan, Ö. 2010. *Grinding Process, and Cement Fineness on the Physical and Mechanical Properties of GGBFS-, Natural Pozzolan-, and Limestone-incorporated Cements*, Working Paper AR-GE 2010/01-B, Turkish Cement Manufacturers' Association [TÇMB], 45p. [in Turkish].)

The reduction in water requirement for a constant workability is influenced by the amount of GGBFS, total amount of cementitious material in the mix and the particle size distributions of both the slag and the PC used. Generally, the reduction in water requirement is not more than about 5% (Wainwright, 1986).

A study on the combined effect of GGBFS content and cement fineness of interground and separately ground blended cements has shown that there is not much difference in the water demand of GGBFS-incorporated mortars and control PC mortars (Tokyay et al., 2010). The results obtained are illustrated in Figure 9.11.

9.2.1.4 Effect of SF

The water demand of concrete containing SF is greater than that of the control concrete without SF. The increased water requirement becomes more pronounced as the amount of SF used increases (Carette and Malhotra, 1983; Mehta, 1986a; Scali et al., 1987). This effect is illustrated in Figure 9.12. Increased water demand is basically due to the very high specific surface area of SF.

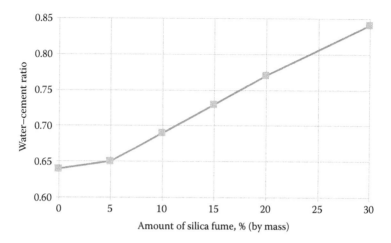

Figure 9.12 Increase in water–cement ratio with increasing SF content, to maintain a given slump of concrete. (Adapted from Carette, G.G., Malhotra, V.M. 1983. Early-age strength development of concrete incorporating fly ash and condensed silica fume, *1st International Conference on Use of Fly Ash, Silica Fume, Slag, and Other Mineral By-products in Concrete, ACI SP-79*, Vol. 2, pp. 765–784.)

9.2.1.5 Effect of limestone powder

Using limestone powder in cement as an addition was found to reduce the water demand by many researchers (Ellerbrock et al., 1985; Sprung and Siebel, 1991; Schmidt, 1992; Opoczky, 1993a,b; Tsivilis et al., 1999; Erdoğdu, 2002; Inan Sezer, 2007) although there are several studies reporting the opposite (Dhir et al., 2007; Tosun et al., 2009). Nevertheless, the change in water demand was not found to be too large. A recent study on the combined effect of limestone powder content and cement fineness of interground and separately ground-blended cements had revealed that among the six blended cements investigated only one had slightly increased water demand (0.4%), whereas the water requirement was reduced in the rest when compared to the control PCs of the same Blaine fineness. Furthermore, the change in water demand was less than 3% for all cements (Tokyay et al., 2010). The results obtained are illustrated in Figure 9.13.

9.2.2 Different workability test results

There is a vast amount of research on the effect of different mineral admixtures on the workability of fresh concretes determined by different test methods. Some of them are summarised in this section.

Superplasticiser requirements of mortars with the same flow and water–cementitious ratio prepared by replacing 30%, 45%, 60% and 75% (by mass) of PC by a GGBFS, a low-lime fly ash and a high-lime fly ash were

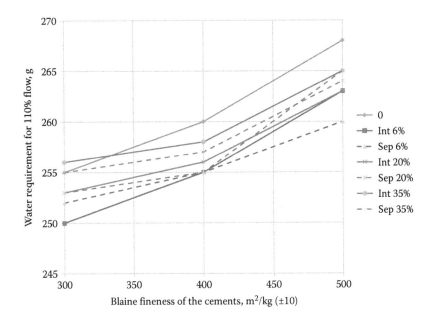

Figure 9.13 Change in water requirement of mortars prepared by using interground and separately ground blended cements with different limestone powder contents and Blaine-specific surface areas. (Adapted from Tokyay, M., Delibaş, T., Aslan, Ö. 2010. *Grinding Process, and Cement Fineness on the Physical and Mechanical Properties of GGBFS-, Natural Pozzolan-, and Limestone-incorporated Cements,* Working Paper AR-GE 2010/01-B, Turkish Cement Manufacturers' Association [TÇMB], 45p. [in Turkish].)

determined. There occurred a considerable reduction in superplasticiser amount up to 30% replacement level for all the three mineral admixtures. The reduction continued for higher replacement levels of high-calcium fly ash. It reached its minimum level at 45% replacement for low-calcium fly ash then there occurred a sharp increase although the superplasticiser dosages were still less than that of the control. Increase in superplasticiser amount was observed in GGBFS-incorporated mortars beyond 30% replacement level, though at a lower rate than that of low-calcium fly ash (Wei et al., 2003).

In a similar investigation, water contents and slump values of the mixes were kept constant for different cementitious material contents. The change in superplasticiser requirement was determined for ternary (PC + GGBFS + FA) and quaternary (PC + GGBFS + FA + one of SF, metakaolin (MK) or limestone powder) mixtures. About half of the PC (by mass) was replaced by mineral admixture combinations. SF and MK constituted 10% whereas the limestone powder constituted 5% of the mineral admixtures used in quaternary mixtures. All the ternary and quaternary mixtures revealed the same or lower superplasticiser requirement than the control mixture. Ternary mixtures were found to result in 23%–43% decrease in superplasticiser requirement. Quaternary mixtures with SF, MK and limestone powder had 0%–23%, 13%–40% and 20%–47% less superplasticiser requirement, respectively (Meddah et al., 2014).

A high-calcium fly ash was found to increase the slump of the concrete for 10%, 20% and 30% cement replacement levels (by mass). The increase in slump was proportional to the increased FA content (Nochaiya et al., 2010). Using additional SF, half the amount of FA, resulted in decrease in slump values although the total cementitious paste volume is increased.

In an experimental investigation on self-compacting mortars, FA, brick powder, limestone powder and kaolinite were used to replace 10% and 20% (by mass) of the PC and three ternary mixtures with 50:50 proportions of FA + limestone powder, brick powder + limestone powder and FA + kaolinite replacing 20% (by mass) of PC were prepared. Changes in flow properties of the mortars were evaluated through the mini slump and mini V-funnel tests described by EFNARC (2002). FA and limestone powder were found to improve the workability whereas brick powder and kaolinite reduced it. Brick powder and FA were coarser than limestone powder and kaolinite. It was concluded that fineness is not the only parameter of a mineral admixture affecting the workability and particle shape and surface characteristics may be more significant (Şahmaran et al., 2006).

In a similar study on SCC, FA, GGBFS, limestone powder, marble powder and basalt powder were used as partial replacement of cement. Slump flow was increased by FA, GGBFS, limestone powder and marble powder. Basalt powder did not change it significantly (Uysal and Sümer, 2011).

GGBFS (10%–25% by mass of PC used) was incorporated in SCC in another research study. Filling ability (slump flow, V-funnel and T50 flow time tests), passing ability (J-ring, U-box tests) and resistance to segregation (modified slump test) were measured. It was determined that the workability of SCC was improved upto 20% GGBFS with an optimum content of 15% (Boukendakdji et al., 2012).

Flowability of SCC was found to increase with increasing natural pozzolan (a basaltic volcanic ash) content (Celik et al., 2014). Slump and slump flow of FA, BFS and SF incorporated lightweight aggregate concretes were investigated to show that FA leads to greater improvement in workability than the other two, when the mineral admixtures are used individually to partially replace the PC. SF, on the other hand, resulted in better control of bleeding and up-floating of lightweight aggregates but workability loss with time was more rapid. BFS improved workability but not as high as the FA (Chen and Liu, 2008).

The influence of SF on the workability was studied through flow table test on mortars. Water–binder and binder–sand ratios were kept constant. It was found that there occurs an increase in flow up to 15% SF which is then followed by a reduction (Rao, 2003). In a similar study, slump of SF-incorporated concretes was determined. Up to 10% replacement of

PC by SF was found to increase the slump with an optimum of 6% (Shi et al., 2002). SF (10% by mass) was used to replace an ordinary PC and a high-slag cement (35% PC + 65% GGBFS). For a constant water–cement ratio of 0.34–0.35, slump of high slag cement was determined to be much higher than that of ordinary PC. SF incorporation increased the slump for both cases, however the increase was more pronounced for OPC + SF mix (Khatri et al., 1995).

9.2.3 Consistency and cohesiveness

The flow of fresh concrete is commonly described by the Bingham model which has two parameters, the yield stress and the coefficient of viscosity (Figure 9.3). The yield stress which indicates the start of flow is directly related with the consistency of fresh concrete. Once the flow starts, the behaviour is governed by the plastic viscosity. The rheological parameters of fresh cementitious systems (pastes, mortars and concretes) are recently being measured by devices called rheometers that relate shear stresses to varying shear rates applied and thus measure the resistance to flow at different shear rate conditions (Banfill et al., 2001). Many different rheometers were designed for this purpose, however their comparison shows that the yield stress and plastic viscosity values obtained were different from each other. On the other hand, it was promising that all the mixtures tested were ranked in the same order by all the rheometers. Therefore, although the absolute values of Bingham constants obtained may be questionable, their relative changes on the differences made in the mix parameters may be used for comparative purposes. Several studies related with the influence of mineral admixtures on yield stress and viscosity of fresh cementitious systems are summarised.

A comparative study that used an ultrafine fly ash (UFFA), a MK, and a SF was carried out to determine the yield stress and viscosity of cement pastes with constant water–cementitious ratio and having different amounts of high-range water reducing agent (HRWRA). The mineral admixture contents ranged from 0% to 16% of cement, replacing the cement by mass; water–cementitious ratios used were 0.28–0.35 and the dosage of a naphtalene sulphonate based HRWRA was 0.45%–0.70% of the cementitious material, by mass. Replacement of cement by UFFA has led to reduced HRWRA dosage over the control for a given yield stress and viscosity, whereas SF increased it significantly. MK was found to have almost no change in the rheological properties (Ferraris et al., 2001). In the same study, influence of particle size of mineral admixtures on the yield stress and viscosity was determined by using four fly ashes with mean particle diameters of 18 µm (coarse), 10.9 µm (medium), 5.7 µm (fine) and 3.1 µm (ultrafine). All tests were conducted at the same mineral admixture content (12%), same water–cementitious ratio (0.35) and same HRWRA dosage (0.45%). UFFA resulted in the lowest yield stress and viscosity. The highest yield stress was observed in the fine FA-incorporated paste and highest viscosity was observed in the medium FA-incorporated paste.

Matrix mortars of concretes were prepared by using a PC (control) and two blended cements that contain 30% (by mass of cement) replacement of PC by a phosphorus slag and a limestone powder. SF (3%–12% by mass) was further used to replace the cements. Equal amounts of phosphorus slag or limestone powder were reduced in blended cements on SF incorporation. Water, HRWRA and aggregate contents were kept constant in all mixtures. Yield stress and viscosity of blended cement mortars were found to be significantly lower than those of the control. Partial replacement of PC by SF reduced the viscosity as the SF amount used increased up to 9%. At 12% replacement, there occurred a sharp increase in viscosity. A similar trend was observed for the yield stress but the minimum value was obtained at 6% replacement. Yield stress increased at 9% SF content, although it was still less than that of the control. At 12%, it was considerably higher than the control. Neither the viscosities nor the yield stresses of SF-incorporated blended cement mortars changed much with respect to non-SF blended cement mortars up to 9% SF content. 12% SF resulted in increased values of these. The increase in viscosity was higher for the phosphorus slag-blended cement, whereas the increase in yield stress was higher for the limestone blended cement (Shi et al., 2002).

In another study on SF- and FA-incorporated cement pastes, both the viscosity and the yield stress were determined to increase with the increased amount of SF and decrease with the increased amount of FA for a constant water–cementitious ratio (Nanthagopalan et al., 2008).

The flow resistance of SF-incorporated pastes was found to be dependent on the type of plasticising agent used. Vikan and Justnes (2007) used 0%–13.6% (by volume) SF to partially replace the PC to prepare pastes with a constant total solid particle volume fraction of 0.442. A sodium naphthalene sulphonate-formaldehyde (SNF) and a polyacrylate (PA) superplasticiser were used at constant dosages of 1.32% and 0.79% (by mass of total cementitious material) in all mixes. The flow resistance was increased with increasing SF content when SNF was used but decreased when PA was used. On the other hand, limestone powder was found to decrease the flow resistance in the same study.

Laskar and Talukdar (2008) partially replaced the cement by different amounts of FA, SF and rice husk ash (RHA) in three groups of concretes. The mix proportions were held constant in each group. The first and the second had a polycarboxylic ether based and the third had a sulphonated naphtalene-based superplasticiser. Their experimental results are shown in Figure 9.14.

FA resulted in a small amount of decrease of yield stress at around 10%–20% replacement levels. Beyond those values, the change was insignificant. There seems to be slight increase in viscosity for 10% FA incorporation, however, the change is not significant. Viscosity was increased up to 10% replacement by SF and then decreased. Yield stress was also decreased

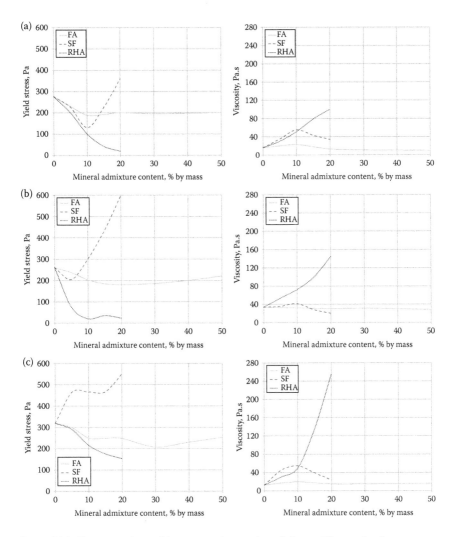

Figure 9.14 Change in the yield stress and viscosity of three different fresh concretes by mineral admixture use. Total cementitious material contents were (a) 563 kg/m³, (b) 518 kg/m³ and (c) 485 kg/m³. First two concretes had polycarboxylic ether polymer and the third had sulphonated naphtalene polymer as high-range water reducing agents. (Data from Laskar, A.I., Talukdar, S. 2008. *Construction and Building Materials*, 22, 2345–2354.)

considerably up to 10% in the mixes with polycarboxylic ether superplasticiser. At higher replacement levels, it increased steeply. In the sulphonated naphtalene superplasticiser used concrete, yield stress increase with increasing SF content although it seems to stay almost the same between 5% and 15% replacement levels. RHA was very effective in reducing the yield stress. On the other hand, it resulted in tremendous increase in viscosity.

9.2.4 Segregation and bleeding

Separation of the fresh concrete components that results in a non-homogeneous mix is called as segregation. Generally, it means the separation of coarse aggregate from the mortar. However, bleeding which may be defined as the appearance of water and some fine particles on the surface of the concrete may be considered as a special form of segregation. It occurs due to the separation of water from the rest of the concrete mix. Usually, these two phenomena are simultaneous.

The factors that increase the possibility of segregation are using (1) larger maximum aggregate size (>25 mm), (2) a high proportion of coarse aggregate, (3) coarse aggregate that has considerably higher density as compared to the fine aggregate, (4) insufficient amount of fine particles (sand and cementitious materials), (5) irregularly shaped and rough surface aggregate particles and (6) too wet or too dry mixes (Mindess and Young, 1981).

While the larger aggregate particles settle, some of the mixing water moves upwards carrying some fine cement and sand particles together causing a layer of scum at the top surface which becomes weak both in terms of strength and durability due to its higher water–cement ratio and porosity. Sometimes, some of the water moving upwards may get collected underneath the coarse aggregate particles and the reinforcement bars as 'water pockets' causing reduced bond of them with the surrounding matrix. A small amount of bleeding may be beneficial since it reduces the water–cement ratio as long as the rate of evaporation from the surface is approximately equal to the rate of bleeding. However, this is seldom the case and if the rate of evaporation is greater than the rate of bleeding, plastic shrinkage cracks may occur.

Using coarse cementitious materials or too much water in the concrete mix are the two basic factors that increase the tendency for bleeding. However, in order to reduce the risk of segregation and bleeding, proper handling, placing, compaction and finishing methods for fresh concrete which are described in most standards and specifications are as important as or even more important than the measures that should be taken for the aforementioned factors.

Properly proportioned concretes that contain carefully selected mineral admixtures will be less prone to segregation and bleeding. Mineral admixtures which are finer than the PC used and which are not composed of flat and elongated particles are preferable. Although use of such materials result in greater plasticity and better cohesiveness, it is necessary to note several peculiar points related with some of the individual mineral admixtures.

When the ratio of surface area of solid particles-to-volume of water is low, the rate of bleeding increases. For example, coarsely ground pumicite (a natural pozzolan) may increase the water requirement of concrete for a given slump and this may lead to increased bleeding and segregation (ACI 232.1R, 2001). On the other hand, for a proper workability, the amount of solid particles must be maximised and the amount of water should be minimised in the paste which means that the mineral admixture used should not be extremely fine unless a HRWRA is used. SF-incorporated concretes do not segregate appreciably both due to the high fineness of the material and the use of high-range water reducing agents. High specific surface area of SF results in significantly reduced bleeding because there remains very little free water in the mixture. Furthermore, bleeding channels and pores are blocked physically by SF particles (ACI 234, 2000). This blocking effect is also true for other mineral admixtures that are finer than PC.

Since SF concretes show much less bleeding, they have the tendency of plastic shrinkage. Therefore, necessary precautions should be taken to prevent the evaporation of moisture from concrete at the early ages especially under conditions such as high fresh concrete temperature, low humidity and high wind (ACI 234, 2000).

The spherical shape and the hydrophilic nature of the low-lime fly ash particles result in a very thin layer of water adsorbed on their surfaces, which leads to an even distribution of the mixing water throughout the fresh concrete. Besides providing a greater surface area of solid particles and requiring lower water contents, low-lime fly ashes reduce bleeding further due to this physicochemical effect, also (Dhir, 1986).

Bleeding capacity and rate of GGBFS-incorporated concretes depend on the fineness of the slag as compared to that of the PC together with which it is used. Finer slag, when it replaces the PC on equal mass basis, results in reduced bleeding whereas coarser slag will cause more bleeding at a higher rate (ACI 233, 2003).

9.2.5 Slump loss

The consistency of fresh concrete changes with time due to the partial consumption of mixing water through absorption by the aggregates, cement hydration or evaporation. These factors may cause a reduction in slump value from the time of initial batching to the time of delivery. It may be an important consideration for the proper placing, consolidation and finishing of fresh concrete.

It was explained in Section 8.3 that mineral admixture incorporation enhances the hydration of cement, however, this effect is suppressed if the amount of mineral admixture used is high (dilution effect). Therefore, the rate of slump loss may be higher in fresh concretes containing small amounts (<10%) of mineral admixtures and lower in those with higher mineral admixture contents. In an experimental investigation, the slump values of a control concrete without any mineral admixture and four different concretes that contained 7% (replacing PC), 20% and 30% (addition, by mass of cement) FA and 10% (addition, by mass of cement) SF were measured at 0, 30 and 60 min after mixing. Initial slumps of the concretes were set at 200 ± 10 mm. Cement replacement by 7% FA resulted in the

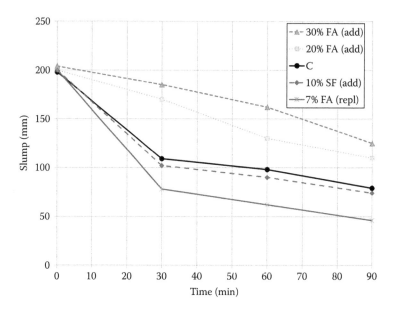

Figure 9.15 Slump loss of FA- and SF-incorporated concretes. (Data from Erdoğdu, Ş., Arslantürk, C., Kurbetci, Ş. 2011. *Construction and Building Materials*, 25, 1277–1281.)

highest rate of slump loss within the first 30 min. The slump loss rates of the control concrete and 10% SF added concrete were nearly the same. Higher FA contents revealed lower rates of slump loss as shown in Figure 9.15 (Erdoğdu et al., 2011).

When it becomes necessary to use water reducing agents or high-range water reducing agents together with mineral admixtures such as SF, the change in slump loss is actually due to the effect of the chemical admixture used (ACI 234, 2000). Different chemical admixtures may result in different rates of slump loss. Therefore, it is advisable to prepare trial batches to determine the slump loss.

9.3 CHANGES IN SETTING TIME UPON MINERAL ADMIXTURE INCORPORATION

Setting is defined as the condition reached by a cement paste, mortar or concrete when it starts to lose its plasticity (ACI 116, 2005). It is different from hardening which describes a condition of having a measurable strength. Cement starts to hydrate as soon as it contacts with water and setting and hardening are the consequences of the continuing hydration process. The initial and final settings are the two arbitrary points at which the paste, mortar or concrete starts to stiffen and attain significant rigidity, respectively. Setting time is measured by penetration resistance tests. The significance of initial and final setting times lies in the fact that the former gives the limit of handling (mixing, placing and compaction) of the fresh cementitious system and the latter indicates the start of gaining strength at a significant rate.

The factors affecting the setting time are (1) amount of water in the mix, (2) temperature and relative humidity, (3) cement properties such as the chemical composition and fineness and (4) type and amount of admixtures in the mix.

The time of setting of hydraulic cement is commonly determined by using Vicat apparatus. The test is described in the international standard such as EN 196-3 (2002) and ASTM C 191 (2008). A similar but less common method which uses Gillmore apparatus is described in ASTM C 266 (2013). The setting time of concrete is determined by a device known as concrete penetrometer and the test is described in ASTM C 403 (2008). Although the setting time of concrete is related to the setting time of the cement that is used, the two values are not the same simply because the setting time of cement is determined by pastes of standard water content that brings the paste to a specific consistency known as normal consistency, whereas the water content of the concrete is usually different from the water content of the normal consistency cement.

Since setting is a consequence of the hydration process, the discussion on the effect of mineral admixtures on the early hydration process also holds true for their effects on the setting behaviour. Nevertheless, the results of several investigations specifically carried out to determine the effects of different mineral admixtures on setting time are discussed next.

A natural pozzolan, a granulated BFS and a limestone were used with the same clinker to prepare CEM II/A-P, CEM II/A-S and CEM II/A-L cements with 6% and 20% (by mass) mineral admixture and CEM II/B-P, CEM II/B-S and CEM II/B-L cements with 35% (by mass) mineral admixture. The cements were prepared by (1) intergrinding and (2) separate grinding and then blending. P and S cements were divided into two fineness groups as 300 ± 10 and 600 ± 10 m²/kg (Blaine-specific surface area). L cements were grouped into 300 ± 10 and 500 ± 10 m²/kg fineness values. Initial and final setting times of these cements were compared with those of a control PC made from the same clinker (Tokyay et al., 2010). The test results are shown in Figure 9.16.

Two of the mineral admixtures (natural pozzolan and limestone) used in this study were softer and the third (GBFS) was harder than the clinker. Thus, intergrinding resulted in finer mineral admixture portions in natural pozzolan- and limestone-incorporated cements, whereas GBFS-incorporation resulted in finer clinker portion. Therefore, the accelerating effect of the finer mineral admixture particles results in shorter initial setting times almost up to 20% mineral admixture content for relatively coarse interground cements than the control. Shorter initial setting times in

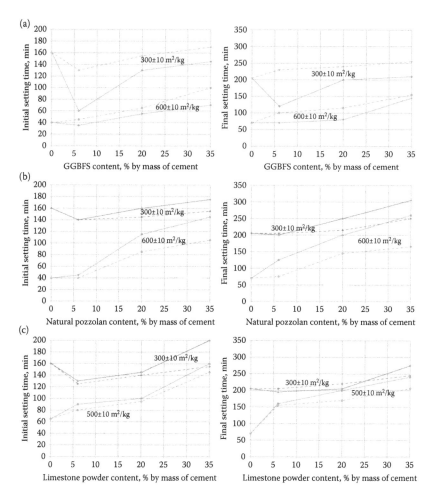

Figure 9.16 (a–c) Change in initial and final setting times of CEM II/A (6% and 20% mineral admixture) and CEM II/B (35% mineral admixture) cements compared with those of the control PCs of the same Blaine fineness. Full lines and dashed lines represent interground and separately ground blended cements. (Adapted from Tokyay, M., Delibaş, T., Aslan, Ö. 2010. *Grinding Process, and Cement Fineness on the Physical and Mechanical Properties of GGBFS-, Natural Pozzolan-, and Limestone-incorporated Cements*, Working Paper AR-GE 2010/01-B, Turkish Cement Manufacturers' Association [TÇMB], 45p. [in Turkish].)

GBFS-incorporated cements is due to the faster reaction of the clinker portion which is finer than that of the control cement because of the additional grinding effect of the GBFS. These effects are not observed in finer cements. Separate grinding and then blending leads to similar reductions in initial setting times for mineral admixtures softer than the clinker. Although there occurs a shortening of initial setting time in GBFS-incorporated separately ground cements, it is not as pronounced as that of interground ones.

In another study, one low-lime and three high-lime fly ashes were used to replace 10%, 20% and 40% (by mass) of the PC. The control PC had 303 m²/kg and the low-lime fly ash (LLFA) 337 m²/kg Blaine-specific surface area. High-lime fly ashes (HLFA1, HLFA2 and HLFA3) had 253, 321 and 970 m²/kg specific surface areas, respectively (Ramyar, 1993). Initial and final setting times determined are given in Figure 9.17.

Figure 9.17 indicates that, regardless of the type of FA, setting times are not significantly affected at 10% replacement level. Actually, fly ashes that are finer than the PC showed slightly less initial setting times than the control PC at this replacement amount. For higher replacement levels, LLFA,

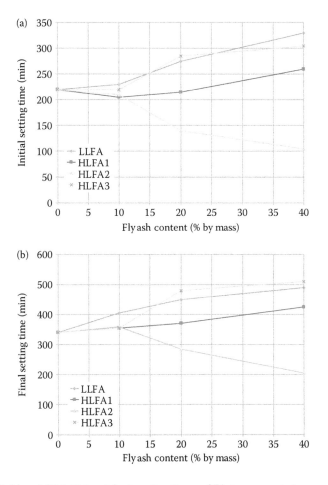

Figure 9.17 (a) and (b) Initial and final setting times of FA-incorporated cements. (Data from Ramyar, K. 1993. Effects of Turkish fly ashes on the portland cement-fly ash systems, PhD thesis, Middle East Technical University, 208pp.)

HLFA1 and HLFA3 resulted in set retardation which is attributed to their higher water requirement for normal consistency. In contrast to the other three fly ashes, HLFA2 accelerated the setting. Mineralogical investigations on this FA had shown that it contains high amounts of free lime, anhydrite and reactive alumina (Tokyay and Hubbard, 1992; Ramyar, 1993). It was observed that anhydrite turns into gypsum on reacting with water and precipitates to result in a false setting. Besides that, gypsum, calcium hydroxide and alumina react with water to rapidly form a considerable amount of ettringite.

In one of the early investigations on the use of SF in concrete it was determined that partial PC replacement by SF does not result in significant changes in setting time although it increases the normal consistency water requirement. When SF was used as 5%, 10% and 20% (by mass) replacement levels, the relative extensions of initial setting time with respect to that of the control PC were 0%, 2% and 8%, respectively (Kayapınar, 1991). These results are confirmed later in a report by the American Concrete Institute (ACI) which states that setting time is not significantly affected by the use of SF by itself (Pietersen et al., 1992).

It is difficult, if not impossible, to generalise the effects of mineral admixtures on the setting time of cementitious systems as numerous factors such as temperature; water content; cement type, source, fineness, composition and amount; use and amount of chemical admixtures; type, amount, fineness, chemical and mineralogical compositions of the mineral admixtures used, etc. influence the setting characteristics. However, if these factors are appropriately considered in the design of cementitious systems, acceptable setting times may be obtained.

9.4 EFFECTS OF MINERAL ADMIXTURE INCORPORATION ON AIR CONTENT OF FRESH CONCRETE

Air voids in concrete are of two kinds. Entrapped air which is present in almost every concrete means the air voids that remain in concrete due to insufficient compaction. Since a full compaction of fresh concrete to attain a no-voids condition is not possible, there is always a small amount of air voids in concrete. This is usually 1%–2%, by volume. It does not exceed 3%. Entrained air is the one that is intentionally introduced into concrete to improve its freezing–thawing resistance. Total air void content of such a concrete is generally 4%–8%, by volume. Entrained air is attained by using air-entraining agents (AEA), which are surfactants having molecules with hydrophilic and hydrophobic ends. These molecules have the tendency to align at the water–air interphase with their hydrophilic ends in water and hydrophobic ends in air. Thus, stable air bubbles form within the cement paste. The entrained air bubbles have sizes between 0.05 and 1.00 mm.

They have spherical shape and are close to each other. Besides the total air content, a critical parameter of air entrainment is the spacing factor which is defined as the average maximum distance from any point in the paste to the edge of the air bubble. The spacing factor should not be more than 0.2 mm for an adequate freeze–thaw resistance (Mindess and Young, 1981). The mechanism will be discussed in more detail in Chapter 11.

The effect of mineral admixtures on the entrapped air is not significant. Generally, a small amount of reduction in entrapped air occurs on mineral admixture incorporation due to the influence of fine particles. On the contrary, for a given dosage of AEA, the total entrained air content decreases with the increasing amount of fine materials in concrete. Such general statements appear in the state-of-the-art reports on the use of different mineral admixtures in concrete (Hjorth et al., 1988; Pietersen et al., 1992; ACI 232.2R, 2003). Among the numerous factors such as fineness and alkali content of the cement; shape, surface texture and the maximum size of the aggregate; chemical admixtures; mixing time; slump; time and frequency of vibration and even pumping and surface finishing; mineral admixtures is just one (Nagi et al., 2007).

The literature on the effect of mineral admixtures on air content of concrete are mostly related with fly ashes. An extensive review of literature on the interaction of FA with air entrainment points out that although there are some unresolved issues within this field, the interaction is definitely related with the carbon content of the FA (Pedersen et al., 2008). As the carbon content increases, the amount of AEA to result in a specified air content also increases. The unburnt carbon that remained in the FA provides adsorption sites for the hydrophobic ends of the AEA molecules thus decreasing the air content. In fact, in an experimental study which also concluded that the carbon content of the FA is the primary variable that determines the air-entraining efficiency, it was determined that no AEA adsorption occurred in 'carbon-free' fly ashes. The unburnt carbon was removed by keeping the FA samples in an oven for 2 hours at 740°C (Külaots et al., 2003). Besides the carbon content, its fineness is another parameter that affects AEA adsorption. Increasing adsorption of AEA was reported for increasing specific surface area of the unburnt carbon in the FA, although the correlation is rather poor (Pedersen et al., 2008).

Some fly ashes were reported to require little or no change in the AEA dosage which is attributed to the type of the water-soluble alkalies present in them. It is generally accepted that as the total amount of alkalies in the FA increases the AEA requirement also increases (von Berg and Kukko, 1991). However, higher concentrations of monovalent alkali cations (Na^+ and K^+) in the FA-incorporated cementitious systems were determined to result in better surfactant behaviour of AEAs, whereas higher concentrations of divalent alkaline earth cations (Ca^{2+} and Mg^{2+}) resulted in reduced AEA efficiency due to the precipitation of AEA from solution on 'hard water' effect (Baltrus and LaCount, 2001; Külaots et al., 2003).

9.5 EFFECT OF MINERAL ADMIXTURES ON EARLY HEAT OF HYDRATION

Unhydrated cement compounds are in a state of high energy since they are formed at very high temperatures in the rotary kiln. On reacting with water, they form products of higher stability and lower energy and the excess energy is given off as heat. Thus, during the setting and hardening processes, the temperature of fresh cementitious systems increases due to the heat liberated on hydration reactions. Both the amount of heat evolution and its rate may be significant in many concrete applications such as mass concrete, hot weather and cold weather concreting and high early strength concretes.

The heat release during the hydration process may be beneficial in some cases and excessive heat evolution may not be desired in other cases. High early heat of hydration is needed in cold weather concreting in order to compensate the retarding effect of low ambient temperature on the setting, hardening and strength gain processes. The opposite is true for hot weather concreting. Concrete being a fair insulator, does not allow the heat developed in its inner portions to be given off readily, whereas the outer portions being in direct contact with the atmosphere will dissipate heat rapidly. Thus, the surface contraction due to cooling is restrained by the expanding interior and there develops tensile stresses that may lead to cracking. These conditions usually result within the early stages of hydration.

Besides these, as it was already discussed in Chapter 8, the heat of hydration and rate of heat of hydration measurements may be used to monitor the extent of hydration of cementitious systems.

The heat of hydration of a PC is approximately the weighted sum of the heats evolved by its individual compounds. Amount of heat evolved at different ages on various hydration reactions of cement compounds are given in Table 9.1. About 50% of the total heat of hydration is liberated in the first 3 days and 70% within 7 days (Mehta and Monteiro, 2006). There occurs intense heat liberation within few minutes as a result of the rapid hydration of C_3S and C_3A during the preinduction period. Sometimes the

Table 9.1 Heat evolution of individual PC compounds upon hydration

Compound	Reacting phases	Reaction product	Heat evolved (J/g)		
			3 days	90 days	Full hydration
C_3S	H	C–S–H + CH	243	435	520
C_2S	H	C–S–H + CH	50	176	260
C_3A	$H + C\bar{S}H_2$	$C_6A\bar{S}_3H_{32}$	887	1302	1670
C_4AF	H + CH	$C_3(A,F)H_6$	289	410	420

Source: Odler, I. 1988. Hydration, setting and hardening of portland cement, in *Lea's Chemistry of Cement and Concrete*, 4th Ed. (Ed. P.C. Hewlett), pp. 241–289, Elsevier, London; Mehta, P.K. and Monteiro, P.J.M. 2006. *Concrete*, 3rd Ed., McGraw-Hill, New York.

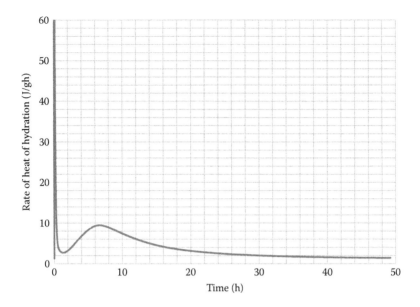

Figure 9.18 Early heat evolution rate of an ordinary PC during hydration.

heat of hydration of plaster of paris or anhydrite which results from the loss of some or all of the structural water of the gypsum rock ground with a rather hot clinker or heat developed in the grinding mill due to interparticle friction may also be added to this first peak. Within the induction (dormant) period, heat evolution drops to a minimum. Then, a second peak appears within 10–16 h of hydration due to the hydration of C_3S resulting in C–S–H and CH. After that, the heat release is slowed down and it becomes very low after a few days. In some of the PCs, a shoulder after the second peak, which is generally attributed to the renewed ettringite formation may be observed in the descending portion of the rate of heat of hydration curve. These features are illustrated in Figure 9.18.

Most mineral admixtures reduce the heat of hydration when they are used to partially replace PC and the reduction becomes more pronounced as the amount of mineral admixture incorporation increases, as shown in Figure 9.19 for a natural pozzolan and GGBFS. In fact, one of the main reasons of using mineral admixtures such as natural pozzolans and low-lime fly ashes is to lower the heat evolution and slow down its rate in the early stages of hydration. However, the literature contains many conflicting reports on the effects of mineral admixtures on heat of hydration.

Mineral admixtures influence the hydration both physically and chemically as discussed in Chapter 8. These effects are reflected in the heat evolution of cementitious materials during hydration. Although the PC dilution effect reduces the heat evolution, the dispersion and nucleation effects and

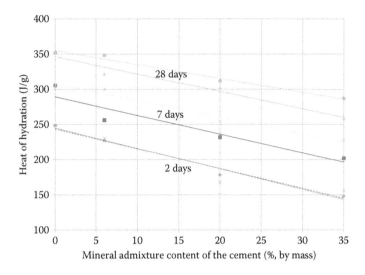

Figure 9.19 Effect of substituting a natural pozzolan (full lines) and GGBFS (dashed lines)
on heat of hydration. (Adapted from Tokyay, M., Delibaş, T., Aslan, Ö. 2010.
*Grinding Process, and Cement Fineness on the Physical and Mechanical Properties
of GGBFS-, Natural Pozzolan-, and Limestone-incorporated Cements*, Working
Paper AR-GE 2010/01-B, Turkish Cement Manufacturers' Association
[TÇMB], 45p. [in Turkish].)

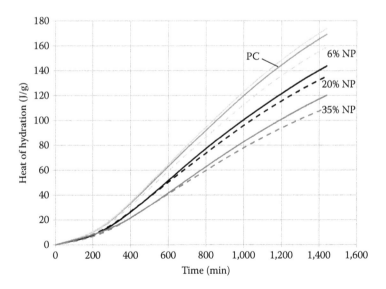

Figure 9.20 Effect of partial PC replacement by natural pozzolan on the early heat of
hydration. Full lines are the measured values and the dashed lines are the
calculated values on the basis of dilution effect, alone. (Data from Ardoğa,
M.K. 2014. Effect of particle size on heat of hydration of pozzolan-incorporated
cements, MS thesis, Middle East Technical University, 109 pp.)

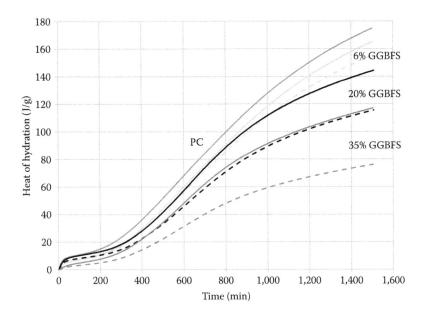

Figure 9.21 Effect of partial PC replacement by GGBFS on the early heat of hydration. Full lines are the measured values and the dashed lines are the calculated values on the basis of dilution effect, alone. (Data from Çetin, C. 2013. Early heat evolution of different-sized portland cements incorporating ground granulated blast furnace slag, MS thesis, Middle East Technical University, 109 pp.)

modification of particle size distribution by mineral admixture incorporation may result in increased early heat due to accelerated hydration. Thus, the resulting heat of hydration is usually higher than that is expected based on the dilution effect alone.

The accelerating effects may dominate the dilution effect for small amounts of PC replacement by mineral admixtures and the early heat evolution may even be higher than that of the control cement. Besides these, many other properties such as fineness, chemical composition, surface characteristics and alkali release rate of the mineral admixture used would be effective in changing the hydration and thus the heat evolution rate of the cement. The early heats of hydration measured and calculated on the basis of dilution effect alone for different amounts of a natural pozzolan and a GGBFS partially replacing the PC are graphically shown in Figures 9.20 and 9.21, respectively.

Chapter 10

Effects of mineral admixtures on the properties of hardened concrete

The properties of concrete in the hardened state can be discussed in three broad categories as strength, dimensional stability and durability. These groups are generally interrelated with each other and each is comprised of several subgroups. Furthermore, all the properties of hardened concrete depend on its properties and the conditions encountered in the fresh state and at early ages.

Mineral admixture incorporation in concrete results in changes in almost all of the properties of hardened concrete as compared to those of concrete without any mineral admixtures. The effects of mineral admixtures on the strength, dimensional stability and durability of concrete will be discussed in the following sections. In order to understand these effects, the basics related with each of these three categories will be given at the beginning of each section.

10.1 STRENGTH

Strength is usually considered as the most important property of concrete which indicates its quality. It is generally accepted that any improvement in concrete strength would correspond to improvements in other hardened concrete properties. It should be kept in mind that although this happens to be true for most cases, there are quite a number of exceptions. Certain steps taken to enhance the strength may lead to dimensional stability or durability problems. Nevertheless, concrete strength (usually the compressive strength) is universally taken as an index property. One other main reason for this is the ease of measuring the compressive strength in comparison with the other properties of hardened concrete.

10.1.1 Basics of strength of concrete

Concrete is a complex material. Although only two phases which are the aggregate particles of various sizes and shapes and the binding cementitious medium (the cement paste) are visible to a naked eye, neither of these two phases are homogeneously distributed within the concrete body nor they are homogeneous in themselves. Aggregates contain several different mineral phases, voids and even microcracks; the cement paste is composed of various hydration products, unhydrated cement particles, pores and voids of different sizes and nature. Furthermore, in the presence of aggregate particles, the paste close to them differs from the bulk of the paste. A thin layer of 10–50 µm thickness which is called the interfacial transition zone (ITZ) appears around aggregate particles. This zone is weaker than the other two phases because it contains more and larger voids and ettringite and calcium hydroxide crystals than those in the paste away from the aggregates. In fact, ITZ may be considered as a third phase in concrete and is the weakest link from the strength point of view.

The strength of concrete therefore, must be related with the properties of these three phases and their interactions. However, the nature of forces between them or even within any of the individual phases is not fully understood, yet. Furthermore, the microstructures of the cement paste and the ITZ change with time, temperature and humidity. Thus, with so many complexities, it is very difficult to establish strength models for concrete.

On the other hand, as with every solid material, there exists an inverse relationship between porosity and strength in concrete. In other words, as the porosity of concrete increases, strength decreases. Although this generalisation oversimplifies the system by not considering the nature of the material, sizes and shapes of the pores, whether the pores are empty or filled with water and if the pores are interconnected or isolated, it is still evident that strength is mainly dependent on porosity.

There had been many attempts to find porosity–strength relationships for cement-based materials. The mathematical models proposed by such studies may be grouped into four as (1) linear, (2) power, (3) exponential and (4) logarithmic, the general equations of which are given, respectively (Kumar and Bhattacharjee, 2003; Lian et al., 2011).

$$\sigma = \sigma_0 - Kp \tag{10.1}$$

$$\sigma = \sigma_0(1 - p)^m \tag{10.2}$$

$$\sigma = \sigma_0 e^{-Kp} \tag{10.3}$$

$$\sigma = K \ln\left(\frac{p_{0s}}{p}\right) \tag{10.4}$$

where,
   σ is the compressive strength at porosity, p
   σ_0 the compressive strength at zero porosity
   p_{0s} the porosity at zero strength
   m and K are the empirical constants

Besides these general equations, there are several other more sophisticated models such as those proposed by Odler and Rössler (1985); Atzeni et al. (1987); Luping (1986) and Kumar and Bhattacharjee (2003).

Beyond all these, the classical work of Powers (1958) that related the compressive strength of PC mortars to their capillary porosity is still very valuable. The porosity of any cementitious system may be grouped into two as (1) gel porosity and (2) capillary porosity. Gel porosity is an integral part of the gel formed on the hydration of cement and these pores have sizes smaller than 10 nm. Gel porosity is constant for cements and it is 0.26–0.27 per gram of cement. Since the specific surface areas of the hydration products are very large, the secondary forces between them are also sufficiently large. Therefore, gel pores do not affect the strength. On the other hand, capillary porosity is an important parameter that affects strength. They, together with microcracks which are always present in the cement paste, are responsible for the stress concentrations and the start of failure. Capillary pores may have sizes in the range of 10–10,000 nm. The volume of the capillary pores and voids in a cement paste depends on (1) the degree of hydration and (2) the amount of mixing water. These features are illustrated in Figures 10.1 and 10.2, respectively.

Powers had made use of these to define a quantity that he named as gel–space ratio which is the ratio of the volume of hydration products, including the gel pores to the total space available. The total space available is the sum of the volume of the hydration products and capillary pores (Powers, 1958). The gel–space ratio can be thought of as a measure

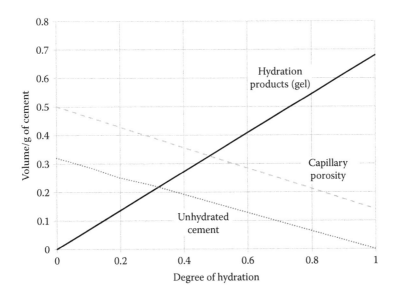

Figure 10.1 Relative volumes of hydration products (gel), capillary porosity and unhydrated cement for a PC paste with water–cement ratio of 0.50.

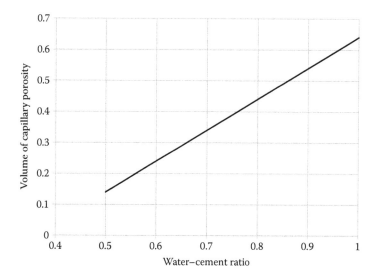

Figure 10.2 Change in the capillary porosity with water–cement ratio for a fully hydrated PC paste. Note that, full hydration is not possible for water–cement ratios less than 0.42. (Adapted from Mindess, S., Young, J.F. 1981. *Concrete,* Prentice-Hall, Englewood Cliffs, NJ.)

of the compactness of a cement paste. It is calculated by Equation 10.5, given here.

$$X = \frac{\text{volume of gel (including gel pores)}}{\text{volume of gel (including gel pores)} + \text{volume of capillary pores}}$$

$$X = \frac{0.68\alpha}{0.32\alpha + (W/C)}$$

(10.5)

where
 X is the gel–space ratio
 α the degree of hydration
 W/C the water–cement ratio

The relationship between the gel–space ratio and strength was proposed as

$$\sigma = aX^3$$

(10.6)

From the results of an extensive experimental study which used 5 cm mortar cube specimens. The coefficient 'a' in the equation is sometimes named the intrinsic strength (strength at zero porosity) which is not completely correct since it only considers the capillary porosity. On the other hand, it is not of much advantage to include the gel porosity in the calculation since it is always a constant fraction of the gel. The value of the coefficient 'a' is 230–235 MPa.

It should be stated once more that Powers' equation holds true for mortars. Porosity and strength can be related to each other in cement pastes and mortars. However, the situation in concrete is far too complex for making such predictions.

10.1.2 Factors affecting the strength of concrete

On page 2 of his classical work *Design of Concrete Mixtures* Abrams (1919, p. 2) stated that 'with given concrete materials and conditions of test the quantity of mixing water used determines the strength of the concrete, so long as the mix is of a workable plasticity'. After about 100 years, this statement still holds true for normal concretes. After carrying out more than 50,000 tests, most of which were compressive strength tests on concrete specimens, the following empirical equation relating the compressive strength of concrete to its water–cement ratio was proposed:

$$S = \frac{A}{B^x} \tag{10.7}$$

where S is the compressive strength of concrete and x is the water–cement ratio, by volume. This equation was then rewritten in terms of water–cement ratio, by mass as follows:

$$S = \frac{A}{B^{1.5(w/c)}} \tag{10.8}$$

The empirical constant A is usually taken as 97 MPa and the constant B depends on parameters such as the type of cement, age of concrete, curing conditions, etc. B may be taken to be about 4 (Abrams, 1919; Mindess and Young, 1981).

Since the water–cement ratio determines the porosity of the cement paste, it is obvious that the concept of gel–space ratio discussed in the previous section forms the basis of Abram's 'Law'. On the other hand, the law holds true for a given set of concrete ingredients and conditions. When the water–cement ratio is kept constant, there are many other factors affecting the strength.

Generally, the factors affecting the strength of concrete are grouped into four as (1) strengths of the component phases, (2) curing conditions, (3) specimen parameters and (4) loading parameters.

10.1.2.1 Strengths of the component phases

Concrete is composed of three phases as (1) the matrix (cement paste) which is the binding phase, (2) the particulate matter phase (aggregates) and (3) the ITZ between the cement paste and the aggregate.

In the process of making a concrete with desired properties, the first step is the selection of appropriate materials and their suitable relative proportions. Such a selection has influences on the properties of all three phases and most of the time they are interdependent. Besides the three basic ingredients (cement, water and aggregates), concrete usually contains chemical and mineral admixtures. Since the influences of the mineral admixtures on concrete strength will be considered in a much more detailed manner in the following sections, the effects of the first four will be summarised here.

Water: The most important effect of water on concrete strength is directly related with its amount. The higher the water content of the mix, the lower will be the strength. The reduction in strength with increased amount of water is a result of increased porosity of both the matrix and the ITZ due to the higher water–cement ratio. Thus, there is a direct relationship between water–cement ratio and strength for low and normal strength concretes, as shown in Figure 10.3.

The relationship between the water–cement ratio and compressive strength has been found to be valid for higher strength concretes (ACI 363, 1997). However, for concretes with water–cement ratios less than 0.30, much higher strength increases can be attained by reducing the

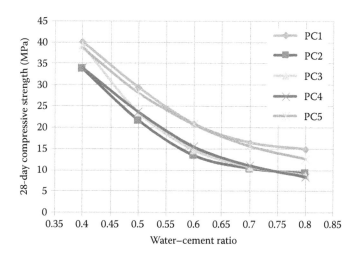

Figure 10.3 Water–cement ratio 28-day compressive strength relationships of normal concretes made by using five different ordinary PCs. (Data from Türkcan, M. 1971. Concrete strength and water–cement relations for various types of western turkish portland cements, MS thesis, Middle East Technical University, 76pp.)

water–cement ratio. This phenomenon is attributed to the improvement of the ITZ by reduced water content which leads to higher ITZ strength and the increase in the specific surface areas of the hydration products which improves the matrix strength (Mehta and Monteiro, 2006).

Another factor related with the mixing water which may affect the strength of concrete is the presence of impurities. If the mixing water is potable, it can be used without testing or qualification. On the other hand, in case of any doubt about the water, the best way is to test it from the setting time and strength points of view by making a concrete and comparing the results with those of the control concrete with potable or distilled water. The 7-day compressive strength attained with the questionable water should not be less than 90% of and the setting time should not be shortened more than 1 h or elongated more than 1.5 h than that of the control concrete (ASTM C 94, 2000).

Air Content: Air voids are always present in concrete either as entrapped air or entrained air. Usually, entrapped air is not more than 1% of the total volume of a sufficiently consolidated concrete. However, insufficient consolidation may result in higher air volumes. Entrained air, on the other hand, is the air introduced into concrete on purpose by using air-entraining admixtures to improve the freeze–thaw resistance. The amount of entrained air in such concretes is generally 4%–8%, by volume.

Although, it is the water–cement ratio that determines the porosity of the cement paste in concrete, the presence of air voids also have the effect of increasing it and therefore decreasing the strength. Typically, other things being the same, 1% increase in air content would lead to about 5% decrease in the strength of concrete. However, entrained air has the effect of improving the workability of fresh concrete, especially in lean mixes which are usually harsh. Thus, for the same workability, water content may be reduced to compensate some of the strength loss caused by air entrainment.

Cement: The effect of cement on the strength of concrete depends on the compound composition and fineness. Among the two calcium silicate compounds that make up about 70% of most PCs, C_3S is responsible mostly from the early strength and C_2S is responsible from the late strength. It was already discussed in Chapter 8 that the compound composition is one of the major factors that affect rate of hydration of PCs. Therefore, PCs with higher C_3S content gain strength more rapidly since it is the most rapidly hydrating compound. However, such cements may have slightly lower ultimate strength. The compressive strengths at 2, 7, 28 and 90 days of five laboratory-made PCs with different C_3S contents are shown in Figure 10.4.

Cement fineness is another important factor affecting the rate of hydration and therefore the strength development. For a given compound composition, the finer the cement, the more rapid will be the hydration. It is generally accepted that cement particles larger than 45 μm hydrate slowly and those which are larger than 75 μm may never hydrate completely (Mehta and Monteiro, 2006).

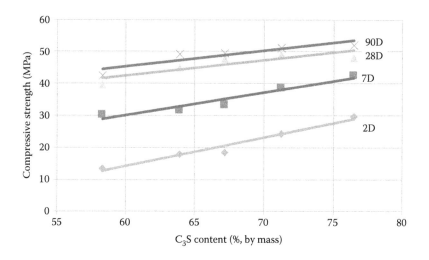

Figure 10.4 Compressive strengths of PCs with different C₃S contents, at different ages. (Data from Yılmaz, A. 1998. Effect of clinker composition on the properties of pozzolanic cements, MS thesis, Middle East Technical University, Ankara, Turkey, 103pp).

Cement particles usually range from 0.5 to 150 μm. Strength development of any PC mainly depends on the particles with sizes between 3 and 30 μm. It is recommended that the amount of particles within this size range be 40%–50% for normal strength cements and 55%–70% for high early strength cements. Particles smaller than 3 μm react with water very rapidly

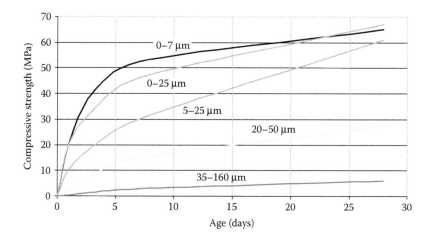

Figure 10.5 Compressive strength development of different size ranges of the same PC. (Adapted from Duda, W.H. 1977. *Cement Data-Book*, 2nd Ed., Bauverlag GmbH, Weisbaden.)

and are completely hydrated before 24 h (Duda, 1977). The strength development of the same PC with different size ranges is given in Figure 10.5.

In a more recent study (Ardoğa, 2014) on the effect of particle size of cements on the heat of hydration, a PC was separated into different size ranges as <10, 10–35, 35–50 and >50 µm by using a sonic sifter. The rate of early heat evolution of each size group was measured. It was found that cement particles smaller than 10 µm hydrate much more rapidly than original cement, whereas all other particle size ranges resulted in lower rates of hydration and as the particles get coarser, the rate of heat evolution became smaller as illustrated in Figure 10.6. This behaviour is generally attributed to the availability of higher specific surface area of the finer particles for hydration. However, it should also be noted that the compound compositions of different size ranges differ because the clinker compounds have different grindabilities. For example, for the same grinding effort, C_3S gets finer than C_2S (Duda, 1977). Indeed, in several investigations including the study stated earlier, it was found that C_3S contents of the finer portions of the PCs were higher than those of the original cement and the coarser portions (Tokyay et al., 2010; Över, 2012; Ardoğa, 2014; Çetin, 2014).

Aggregate: Strength of aggregate is usually not a critical factor affecting the strength of normal concrete. Except for lightweight aggregate concretes and some high strength concretes, full strength of aggregate is

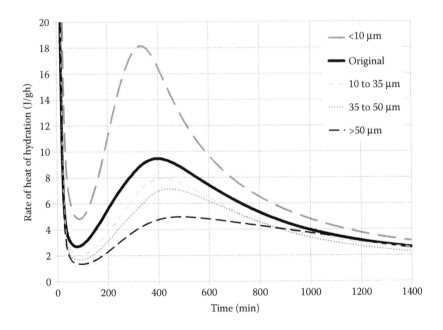

Figure 10.6 Rate of early heat evolution of different size ranges of the same PC. (Data from Ardoğa, M.K. 2014. Effect of particle size on heat of hydration of pozzolan-incorporated cements, MS thesis, Middle East Technical University, 109pp.)

seldom utilised because the strengths of normal weight and heavy weight aggregates are more than those of the matrix phase and the ITZ. However, there are certain properties of aggregate such as size, shape, surface texture, grading, etc. which affect the concrete strength.

Generally speaking, the effect of aggregate properties on concrete strength are due to the changes they result in the water–cement ratio of the concrete although several aggregate properties like maximum aggregate size, surface texture and mineralogy would affect the properties of the ITZ to a certain degree.

Concrete aggregates are almost always required to have a continuous grading that is, a concrete aggregate should contain fine, medium and coarse particles in appropriate relative proportions in order to minimise the voids content. Continuous grading is also important for workability and economical aspects. All standards on concrete aggregates set limits for aggregate gradings which are based on practical experience.

Maximum aggregate size, which is defined as the smallest sieve opening through which the entire aggregate sample passes significantly affects the concrete strength. Usually, job conditions such as the smallest dimension of a concrete member or the clear spacing between the reinforcement bars enforce the maximum aggregate size. The larger the maximum size of a well graded aggregate, the less will be the water–cement ratio of the concrete mix for a given cement content and concrete workability, since the surface area to be wetted will decrease as the aggregate gets larger. Thus, the strength increases. The water-reducing effect of increased maximum aggregate size is more pronounced in lean concrete mixtures in which the cement contents are lower than 300 kg/m^3 of concrete (Mindess and Young, 1981). On the other hand, decreasing surface area due to increased aggregate size also leads to a reduced bond between the aggregate and the matrix at ITZ (Akçaoğlu et al., 2002, 2004). This latter effect becomes more noticeable as the concrete strength increases. It can thus be concluded that larger aggregates require less water, however, they form weaker ITZ. Because ITZ plays a more important role in stronger concretes, smaller maximum aggregate sizes are preferable for high-strength concretes.

When concretes are compared on the basis of same workability, the shape or surface texture of the aggregate is not critical. Nevertheless, angular and rough surfaced aggregates improve the mechanical bond with the matrix whereas round and smooth surfaced aggregates reduce the water requirement.

Admixtures: Since the effects of mineral admixtures will be considered in detail in the following sections, only the effects of chemical admixtures will be summarised here. The effects of air-entraining admixtures has been already given. The retarding and accelerating admixtures, as their names imply, may slow down or speed up the strength development to a certain extent at early ages but the ultimate strengths remain almost unchanged. The retardation or acceleration effects of these admixtures are mainly on the setting process. Water-reducing and high-range water reducing admixtures result in increased strength at a given workability due to the lowered water–cement ratios.

10.1.2.2 Curing conditions

Properties of concrete are improved with time as long as conditions are favourable for the hydration of cement to continue. The rate of improvement of almost any desired property of concrete (strength and other mechanical and durability characteristics) is rapid at early ages and continues slowly at later ages. The key to the improvement of strength (and durability) of concrete is the degree of hydration of the cement. However, the rate of pore filling by the hydration products may be as significant as the degree of hydration. In other words, although high water content result in a high degree of hydration, it would be difficult for the hydration products to fill the initially higher volume capillary pores. Therefore, it is more rational to reduce the initial porosity by lowering the water–cement ratio and then take the necessary measures to prevent loss of water from the concrete (ACI 308, 2001). An example of the effect of initial capillary porosity, in terms of water–cement ratio, on the strength development of the same concrete is shown in Figure 10.7.

The primary objective of curing concrete is to keep it saturated or at least, as nearly saturated as possible since hydration takes place in water-filled capillarities. When water is lost from the concrete, so that relative humidity

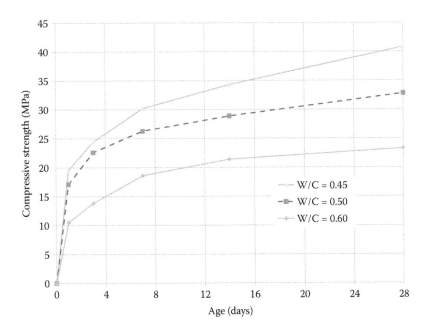

Figure 10.7 Effect of water–cement ratio on the strength development of concrete.

(RH) in the capillary pores reduces below 80%, hydration and strength development stop (Mindess and Young, 1981; Mehta and Monteiro, 2006). Besides humidity, temperature plays an important role in hydration since it has an accelerating effect, as in all other chemical reactions. Thus, the three important aspects of the curing process are time, ambient humidity and temperature. Examples on the effects of ambient RH and temperature on strength development are shown in Figures 10.8 and 10.9, respectively.

Note that, in Figure 10.8, the strength gain at 55% RH within the first few days is similar to that at 100% RH. This is because the water is held in the capillary pores by surface tension and it is available for some time for the hydration of cement. As hydration continues the internal RH decreases. Additional water in the case of fully saturated moist curing will move to the localised areas of lower RH and hydration will continue. Such a movement is much slower in partially saturated curing (Mindess and Young, 1981).

The higher the ambient temperature, the more rapid will be the early strength development and the higher will be the strength up to 28 days for concretes moist cured at constant temperatures between 5°C and 50°C. However, lower curing temperatures will lead to higher ultimate strengths (usually beyond 90 days). This is attributed to the more uniform microstructure of the cement paste at lower curing temperatures than that obtained by higher curing temperatures (Mehta and Monteiro, 2006).

Curing of concrete at elevated temperatures such as steam curing or autoclave curing is primarily used for precast or prefabricated concrete elements although the former may also be applied to enclosed cast-in-place concretes.

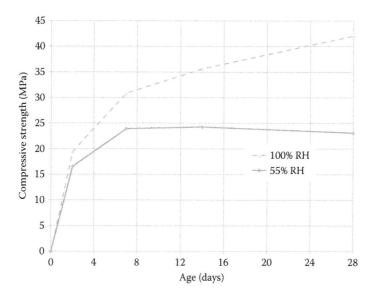

Figure 10.8 Effect of ambient RH on the strength development of PC mortar. (Data from Ün, H., Baradan, B. 2011. *Scientific Research and Essays*, 6(12), 2504–2511.)

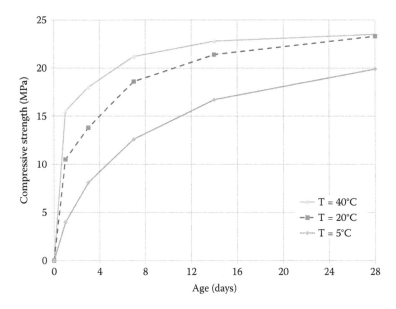

Figure 10.9 Effect of ambient temperature on the strength development of PC concrete. (Data from Kasap, Ö. 2002. Effects of Cement Type on Concrete Maturity, MS thesis, Middle East Technical University, 96pp.)

The main idea is to increase the rate of production by applying shorter periods of curing. Maximum curing temperatures for atmospheric pressure steam curing may range from 40°C to 100°C although the generally used temperature range is 60–80°C. High-pressure steam curing is carried out in autoclaves where the temperature is 150–200°C and the pressure is 5–20 atm.

Time-strength relations for concrete are generally given for moist curing conditions at normal temperatures. Two such relations recommended by ACI 209 (1997) and CEB-FIP (1990) are given next, respectively. The ACI equation is for ordinary PC (ASTM Type I) concrete whereas the CEB-FIP equation may be used for high-early strength and slow-hardening cements as well.

$$\sigma_{ct} = \sigma_{c28}\left(\frac{t}{4 + 0,\,85t}\right) \tag{10.9}$$

$$\sigma_{ct} = \sigma_{c28}\, e^{[s(1-\sqrt{28/t})]} \tag{10.10}$$

where,

σ_{ct} is the compressive strength at t days

σ_{c28} the compressive strength at 28 days

s = 0.20, 0.25 or 0.38 for high-early strength, ordinary portland and slow-hardening cements, respectively

The combined effects of time and curing temperature on concrete strength may be evaluated by using the maturity concept which is simply defined as the product of time and temperature. The idea originated from the early experimental work on steam-cured concrete (Nurse, 1949; Saul, 1951) and the following equations were proposed:

$$M(t) = \sum (T_a - T_0)\Delta t \tag{10.11}$$

$$M(t) = \int_0^t (T_a - T_0)dt \tag{10.12}$$

where
   $M(t)$ is the maturity
   T_a the concrete temperature during time Δt
   T_0 the datum temperature which depends on the type of cement used but generally taken as $-10°C$

These equations brought out the concept of equivalent age (t_e) which allows the estimation of the time of curing at a certain temperature to attain a specified strength with reference to a standard curing temperature (T_r):

$$t_e = \frac{\sum (T_a - T_0)\Delta t}{T_r - T_0} \tag{10.13}$$

Since their appearance in early 1950s, these equations had been a matter of debate but they still find common practical use in spite of their limitations. It is worth to note that the following equivalent age equation of Freiesleben Hansen and Pedersen (1977), which is based on maturity determined by an Arrhenius function is found to be more appropriate:

$$t_e = \sum_0^t e^{-(E_a/R)((1/T)-(1/T_r))}\Delta t \tag{10.14}$$

where
   E_a is the apparent activation energy, J/mol
   R the gas constant (8.314 J/mol°K)
   T the concrete temperature during time Δt, °K
   T_r the reference temperature, °K
   E_a is proposed as 33,500 J/mol for concrete temperatures (T_a) above 20°C and $33500 + 1420(20-T_a)$ for concrete temperatures below 20°C

10.1.2.3 Specimen parameters

Concrete strength tests are sensitive to size, shape and moisture state of the specimens. There are two types of specimens that are almost universally being used for standard compressive strength determinations: cylinder specimens with 15 cm diameter and 30 cm length or 15 cm cube specimens. However, there are occasions such as testing of cores taken from the existing structures or very high strength concretes that would necessitate smaller sized specimens. There have been many investigations on the effect of specimen size and shape for the last 90 years. The general conclusions arrived at may be summarised as (1) larger specimens tend to present lower apparent strengths due to the higher probability of having flaws and (2) cube specimens give higher strengths than equivalent cylinder specimens (it is customary to relate the cylinder strength to cube strength by a multiplication factor of 0.85, for normal strength concretes, but it decreases as the strength of concrete increases).

The standard cylinder specimens have a length–diameter (l/d) ratio of 2.0. If smaller l/d values are used, as would be the case in concrete cores, higher apparent strengths are obtained due to a phenomenon known as the end effect which arises from the friction developed at the top and bottom surfaces of the concrete test specimen. The lateral expansion tendency of the concrete is reduced by the steel platens of the testing machine at the ends of the specimen because steel has a modulus of elasticity about 10 times higher than concrete. The end effect diminishes towards the centre and becomes negligible at the central portion of the specimen which corresponds to about l/d = 1.7. This is also the reason why cube specimens give higher strength values than cylinder specimens made of the same concrete.

10.1.2.4 Loading parameters

Strength of concrete is almost always considered as the compressive strength unless otherwise stated. This, obviously, does not mean that tensile and shear strengths can be ignored because cracking of concrete frequently occurs as the internal tensile stresses developed exceed the tensile strength and under flexural loading failure occurs as a result of the combined effects of tension, compression and shear. Therefore, the type of stress applied may sometimes be an important parameter affecting the strength of concrete.

Tensile strength of concrete is usually taken as 10% of its compressive strength although it may range from 7% to 11%. Tensile-compressive strength ratio of concrete decreases with increasing age and decreasing water–cement ratio and increases with improved ITZ and reduced aggregate size.

Shear strength of concrete was found to be approximately 20% of its compressive strength from the intersection point of failure envelope of the Mohr diagram with the vertical axis (shear stress axis) (Mindess and Young, 1981).

It is generally agreed that as the rate of loading increases, the apparent strength decreases. The rate of stress is specified as 0.23–0.25 MPa/s in most standards on strength and modulus of elasticity determination methods. Loading rates slightly higher or lower than the specified value would result in small differences which are not significant. However, at very high rates of loading approaching to impact, apparent strength increases and at very low rates of loading apparent strength decreases due to the creep effect encountered. The extent of the changes in strength due to very rapid or gradual loading is also related with the properties of the cement paste, ITZ and various aggregate properties such as angularity, size and surface roughness.

10.1.3 Influence of mineral admixtures on concrete strength

Generally speaking, all the factors that were discussed in the previous section are also true for mineral admixture-incorporated concretes. However, there may be some differences between concretes containing mineral admixtures and PC concretes. The extent of these differences is mainly due to the changes that mineral admixtures result in the matrix and ITZ phases of the concrete and they depend on the (1) type, (2) composition, (3) fineness and (4) amount of the mineral admixture, (5) properties of the PC they are used with and (5) the mix proportioning method employed.

The effects of mineral admixtures on the hydration process and the properties of fresh concrete which consequently affect the strength were already discussed in Chapters 8 and 9, respectively.

10.1.3.1 Strength development

Natural Pozzolans: Strength development in concretes containing natural pozzolans is generally accepted to be slower than that of PC concrete, at early ages. This may not be always the case, especially for lower natural pozzolan contents, as illustrated in Figures 10.10 and 10.11. The former is for concretes with the same consistency (slump = 80–100 mm) and the latter is for mortars with the same water–cementitious ratio. Higher early strengths attained by small amounts of natural pozzolan incorporation may be attributed to the acceleration effect. However, as the amount of natural pozzolan partially replacing the PC increases, the dilution effect becomes more dominant and the strength decreases.

Fly Ashes: The influence of FA on strength development is through its physical effect, pozzolanic reactivity and for the high lime fly ashes, latent hydraulicity. There are a large number of studies reported on the effects of fly ashes on the strength of concrete and there is general agreement among the investigators that the composition and the fineness of the FA are the two most important parameters affecting the strength development

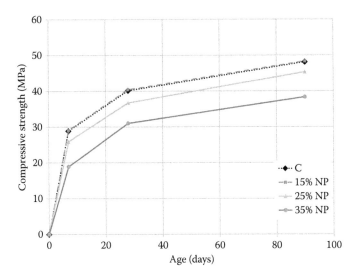

Figure 10.10 Strength development of natural pozzolan-incorporated concretes compared to control PC concrete. PC was partially replaced by 15%, 25% and
35% (by mass) pozzolan and all concretes had the same consistency. (Data
from Shannag, M.J., Yeginobali, A. 1995. *Cement and Concrete Research*,
25(3), 647–657.)

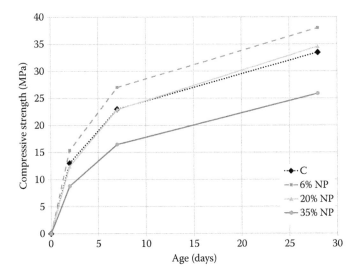

Figure 10.11 Strength development of natural pozzolan-incorporated mortars compared to control PC mortar. PC was partially replaced by 6%, 20% and 35%
(by mass) pozzolan and all mortars had the same water–cementitious ratio
of 0.50. (Data from Delibaş, T. 2012. Effects of granulated blast furnace slag,
trass and limestone fineness on the properties of blended cements, MS thesis, The Graduate School of Natural and Applied Sciences, METU, 72pp.)

in FA-incorporated concretes in comparison with a similar PC concrete. Furthermore, the method of FA incorporation may also be effective.

The effect of FA fineness on the strength development was well described by three investigations (Lawrence et al., 2005; Felekoğlu et al., 2009; Erdoğdu and Türker, 1998). The first one was on a low-lime fly ash and the second one was on a high-lime fly ash. The third one used both a low-lime and a high-lime fly ash. In the first two investigations, original fly ashes were ground to obtain finer samples and mortar specimens were used for strength tests. The study by Lawrence et al. (2005) was a more extensive research on the effects of amount and fineness of ground quartz, limestone powder and two fly ashes. Only the relevant section is included here. The specific surface area (Blaine) and the mean size of the original FA they used were 384 m²/kg and 24 μm, respectively. It was then ground to get three different finenesses as 11, 7 and 5 μm mean particle sizes corresponding to 547, 756 and 909 m²/kg Blaine specific surface areas. PC was replaced by 25% (by mass) fly ashes and water–cementitious ratio was 0.50 in all mixes. The compressive strength development they obtained is shown in Figure 10.12.

A similar study was carried out by Felekoğlu et al. (2009) on a high-lime fly ash that replaced 20% (by mass) PC in the mortars. Instead of a constant water–cementitious ratio, they kept the workability constant as 120 ± 5 mm flow and accordingly adjusted the water content. The mean

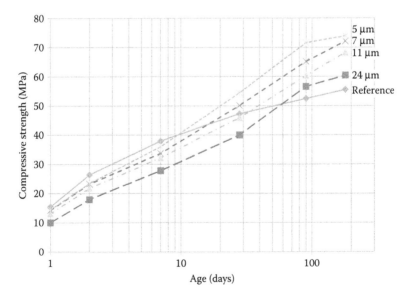

Figure 10.12 Effect of FA fineness on strength development of 75:25 PC-FA mortars. (Data from Lawrence, P., Cyr, M., Ringot, E. 2005. *Cement and Concrete Research*, 35, 1092–1105.)

particle sizes of the fly ashes were 9.87, 7.40, 6.75 and 5.61 μm. All four fly ashes resulted in lower strengths than the control, upto 28 days which was attributed to the increased water requirements of the FA mortars. Beyond 28 days, all FA mortars had higher strengths than the control. However, the sequence was not in the order of fineness as observed in the study by Lawrence et al. The optimum mean particle size of the FA, among the four different sizes investigated, was found to be 7.40 μm.

The third study on the effect of FA fineness on the strength development was different than the previous two studies, in that no grinding was applied for size reduction, instead the original fly ashes were sieved to obtain different size fractions. A high-lime and a low-lime fly ash were used to replace 25% of the PC, in their original form and in size groups of >125, 90–125, 63–90, 45–63 and <45 μm. The strength test results of mortars with constant water–cementitious ratio (0.50) were compared with those of the control PC mortar and mortars with FA incorporated in their original forms (Erdoğdu and Türker, 1998). The strength test results for different size groups are shown in Figure 10.13.

In a comparative study on the strength development in three series of concretes containing silicoaluminous (SA), silicocalcic (SC) and sulphocalcic (SuC) fly ashes, 10% (by mass) PC was replaced by 10%, 20% and 40% FA; and the fine aggregate contents were adjusted to keep the slump values between 60 and 80 mm. The water contents of the mixes were the same as or slightly less than that of the control concrete for direct 10% replacement but higher for the cases of additional FA. SA and SC were conforming with the ASTM C 618 requirements for Class F and Class C fly ashes, respectively. SuC, however, was a non-standard FA. SA and SuC had similar Blaine fineness values around 330 m²/kg whereas SC was coarser (253 m²/kg). The amount of glassy phase was 70%, 50% and 30% in SA, SC and SuC, respectively (Ramyar, 1993). The relative compressive strengths of the FA concretes with respect to the control concrete at different ages are given in Figure 10.14.

Strength development in the low-lime SA and high-lime SC FA incorporated concretes were found to be similar. 10% replacement of PC by all three fly ashes resulted in about 20% reduction in the 7-day strengths. As the hydration and pozzolanic reactions continue, the difference was reduced and concretes with SC had slightly higher strength values beyond 28 days whereas those with SA and SuC had higher strengths beyond 90 days and 180 days, respectively. On using additional FA, SA and SC resulted in strengths higher than those of the control concrete and concrete with 10% replacement, at all ages. However, the use of additional SuC resulted in lower strengths (Ramyar, 1993).

In a similar experimental study, a low-lime fly ash conforming to the requirements of ASTM C 618 was used in different amounts in four groups of concretes with different PC contents. The PC contents of the control concretes were 250, 300, 350 and 400 kg/m³ and the amount of PC was reduced

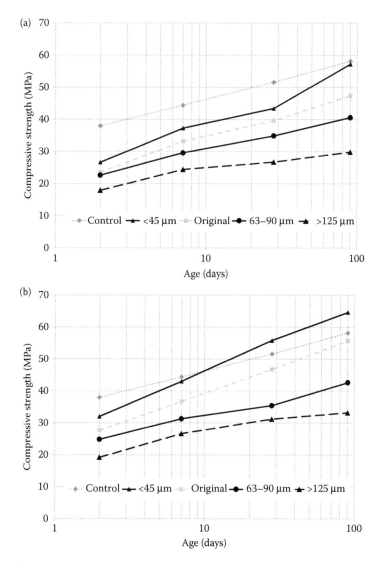

Figure 10.13 Compressive strength development of 75:25 PC-FA mortars containing the original FA itself and coarse (>125 μm), medium (63–90 μm) and fine (<45 μm) size fractions of the fly ashes. (a) Low-lime fly ash, (b) High-lime fly ash. (Data from Erdoğdu, K., Türker, P. 1998. *Cement and Concrete Research*, 28(9), 1217–1222.)

by 20% (by mass) and 15%, 25%, 33%, 42%, 50% and 58% (by mass of the remaining PC) FA was added in the FA concretes. In other words, the ratio of FA to total cementitious material content ranged from 13% to 36.6%. The slumps of all mixtures were kept contant as 120 ± 10 mm and the water and aggregate contents were adjusted (Öner et al., 2005).

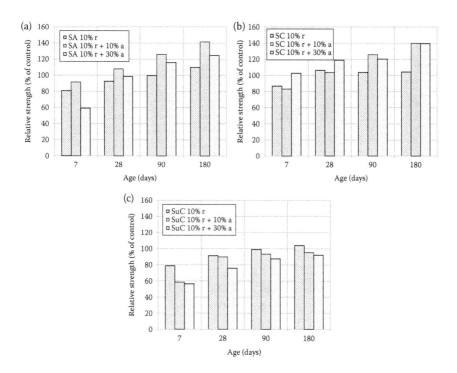

Figure 10.14 Effect of FA type (a) SA, (b) SC and (c) SuC and method of incorporation on the strength development of concrete. 'r' stands for replacement and 'a' stands for addition. (Data from Ramyar, K. 1993. Effects of turkish fly ashes on the portland cement–fly ash systems, PhD thesis, Middle East Technical University, 208pp.)

FA contents less than the reduced amount of cement increased the water–cementitious ratio of the FA concretes with respect to those of their control concretes, whereas higher FA contents reduced the water–cementitious ratio linearly with increased amount of FA. The relationship between FA content and change in water–cementitious ratio was the same for all FA concrete series, as shown in Figure 10.15. Although the water–cementitious ratio decreases due to the lubricating effect of FA and increases the amount of paste with increasing FA content, the 28-day strengths were found to increase up to 25%–30% FA–cementitious material ratio. Further increase in the amount of FA reduced the strength, as shown in Figure 10.16.

In another comparative study, two high-lime (HL1 and HL2) and two low-lime (LL1 and LL2) fly ashes were used as partial cement replacement in four different series of concretes. The control mixes were designed for 28-day compressive strengths of 40, 60, 65 and 70 MPa. The second, third and fourth series of concretes contained 0.07% (by mass of the cementitious materials) melamine formaldehyde-based HRWRA. Fly ashes were used to replace 10%, 20% and 40% (by mass) of the PC in FA-incorporated mixes.

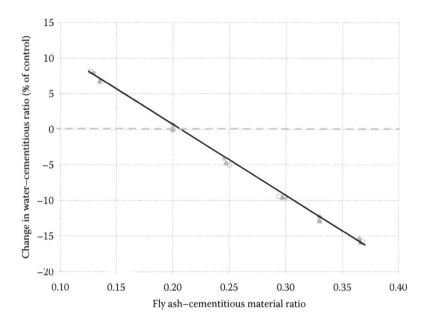

Figure 10.15 Change in water–cementitious ratio of FA concretes with respect to their control concretes. (Data from Öner, A., Akyüz, S., Yıldız, R. 2005. *Cement and Concrete Research*, 35, 1165–1171.)

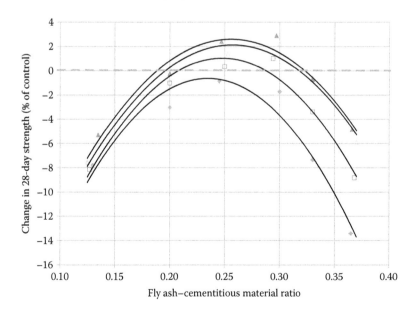

Figure 10.16 Change in 28-day compressive strength of FA concretes with respect to their control concretes. (Data from Öner, A., Akyüz, S., Yıldız, R. 2005. *Cement and Concrete Research*, 35, 1165–1171.)

In all mixes, the slump was kept at 80–100 mm and water contents were adjusted accordingly (Tokyay, 1999). Silica + Alumina + Iron oxide contents of the fly ashes were 73.3%, 72.8%, 89.8% and 91.0% for HL1, HL2, LL1 and LL2, respectively. HL1 had 20.3% and HL2 had 14.0% CaO. LL1, LL2 and HL2 had similar Blaine specific surface areas of 302, 290 and 275 m^2/kg, respectively, whereas HL1 was considerably finer with 455 m^2/kg specific surface area. The effect of each FA on the relative strength is almost the same, irrespective of the mix proportions of the control concretes. Among the two high-lime fly ashes, the finer one was more reactive. The strength reduction with the increased amount of high-lime fly ash was more noticeable at 7 days, at later ages the difference became much less. Among the two low-lime fly ashes, although they had similar chemical compositions and Blaine specific surface areas, LL1 was more reactive. These features are illustrated in Figures 10.17 and 10.18 for two different concrete grades. Comparison of HL1 and LL1 in Figure 10.19 shows that at 10% replacement level they resulted in similar relative strengths that are either slightly higher or the same as the control at all ages tested. Strength reduction was more with the low-lime fly ash as the amount of FA used increased (Tokyay, 1999).

As it can be understood from the foregoing discussion, it is difficult to reach definitive conclusions about the effects of FA properties on the strength of concrete. Nevertheless, some general points can be made:

1. There is only a poor correlation, if any, between the oxide composition of the fly ashes and their effect on the strength development of concretes and it is the mineralogical composition of the FA rather than the chemical composition, that determines the strength development in FA-incorporated concretes (Jawed et al., 1991)
2. For a specified 28-day compressive strength, when FA is used to partially replace the PC, early strengths of low-lime fly ash concretes are lower than those of the control concretes without FA. However, it is possible to attain equivalent early strengths by making proper adjustments in the concrete mix proportions and/or using appropriate chemical admixtures or activators (ACI 232, 2003)
3. High-lime fly ashes are generally more reactive than low-lime fly ashes at early ages (Cook, 1982; Samarin et al., 1983; Tokyay, 1987; ACI 232, 2003); the difference becomes smaller at later ages as the pozzolanic reactions continue
4. As long as the water requirement for a specified workability is not increased significantly, finer fly ashes result in relatively higher strengths

GGBFS: The initial rate of reaction of GGBFS with water is slower than that of PC, which means that when used to partially replace the PC, GGBFS will result in lower early strength. Generally, between 7 and 28 days, strength

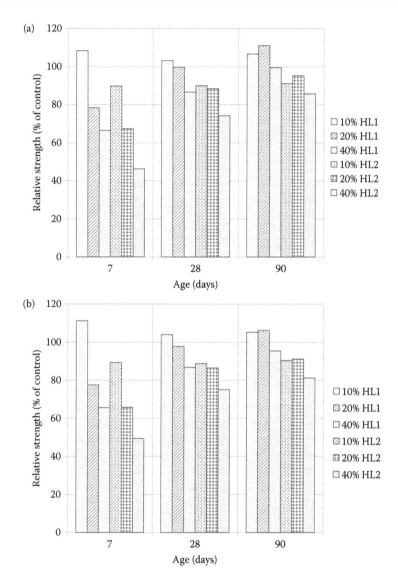

Figure 10.17 Relative strength of 10%, 20% and 40% high-lime fly ash concretes compared to control concretes designed for (a) 40 MPa and (b) 70 MPa 28-days
strength. (Adapted from Tokyay, M. 1999. *Cement and Concrete Research*,
29, 1737–1741.)

approaches to that of the control. The difference between the strength of a
GGBFS-incorporated concrete and a PC concrete may be influenced by many
factors. However, keeping other parameters such as the characteristics of
the PC, time and temperature constant, the reactivity of the GGBFS and its
amount are the two important factors affecting strength development.

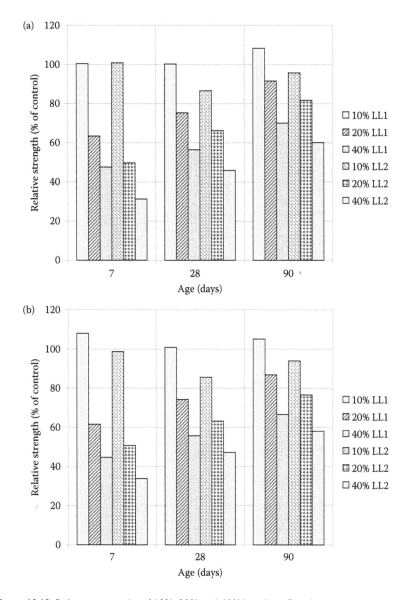

Figure 10.18 Relative strengths of 10%, 20% and 40% low-lime fly ash concretes compared to control concretes designed for (a) 40 MPa and (b) 70 MPa 28-day strength. (Adapted from Tokyay, M. 1999. *Cement and Concrete Research*, 29, 1737–1741.)

The reactivity of GGBFS depends on its chemical and mineralogical compositions and fineness. It is usually specified in terms of the slag activity index (SAI) which is the percent of the strength of 50–50 mass combination of GGBFS-reference PC mortar to the strength of the reference PC mortar, at a specified age. ASTM C 989 specifies three grades of GGBFS as Grade

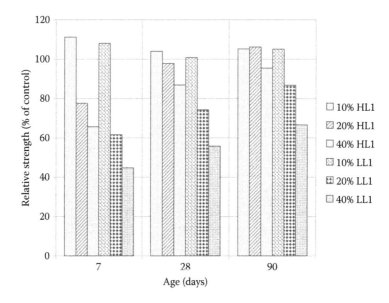

Figure 10.19 Relative strengths of 10%, 20% and 40% high-lime and low-lime fly ashes. (Adapted from Tokyay, M. 1999. *Cement and Concrete Research*, 29, 1737–1741.)

80, Grade 100 and Grade 120 depending on their SAIs, the requirements for which are given in Table 10.1.

Strength tests on mortar specimens prepared according to ASTM C 989 revealed that Grade 120 slag results in slightly lower strengths than PC mortar upto 3 days and higher strength at later ages. Grade 100 slag results

Table 10.1 SAI requirements for the three grades of GGBFS in ASTM C 989 (2014)

	SAI, minimum %	
Grade	Average of last five consecutive samples	Any individual sample
At 7 days		
Grade 80	–	–
Grade 100	75	70
Grade 120	95	90
At 28 days		
Grade 80	75	70
Grade 100	95	90
Grade 120	115	110

lower strengths upto 21 days but higher strengths later. Grade 80 slag how-ever, has lower strength at all ages (ACI 233, 2003).

In an extensive experimental study on the optimum usage of GGBFS in concrete, a GGBFS was used in different amounts in four groups of con-cretes with different PC contents. The PC contents of the control concretes were 250, 300, 350 and 400 kg/m³ and the amount of PC was reduced by 30% (by mass) and 21%, 43%, 71%, 100%, 129% and 157% (by mass of the remaining PC) GGBFS was added in the slag concretes. In other words, the ratio of GGBFS to total cementitious material content ranged from 17.6% to 61.1%. The slumps of all mixtures were kept constant as 120 ± 10 mm and the water and aggregate contents were adjusted, accord-ingly (Öner and Akyüz, 2007).

GGBFS contents less than or equal to the reduced amount of PC increased the water–cementitious ratio of the GGBFS concretes with respect to those of their control concretes whereas higher GGBFS contents decreased the water–cementitious ratio linearly with increased amount of GGBFS. The relationship between GGBFS content and change in water–cementitious ratio was the same for all GGBFS concrete series, as shown in Figure 10.20. The rate of strength development for two concrete series, the control concretes of which were designed for 28-day strengths of 20 and 40 MPa are given in Figure 10.21. All GGBFS concretes except those with 17.6% slag had strengths higher than those of the control concretes at all ages,

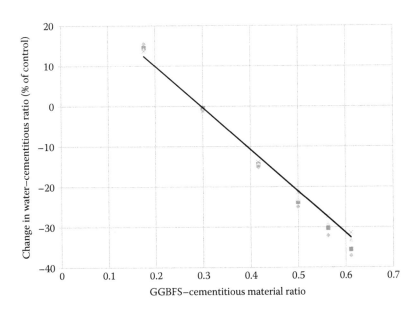

Figure 10.20 Change in water–cementitious ratio of GGBFS concretes with respect to their control concretes. (Data from Öner, A., Akyüz, S. 2007. *Cement and Concrete Composites*, 29, 505–514.)

starting from 7 days. It should be noted that the total cementitious material content was less than those of the control concretes. Concretes with 30% GGBFS represent the one-to-one PC replacement, by mass. The other had more GGBFS than the reduced amount of PC. It is clear from Figure 10.21 that, upto 50% slag content, increase in strength is proportional with the increased amount of slag in the total cementitious material. However, beyond 50%, strength reduces although it is still higher than the control.

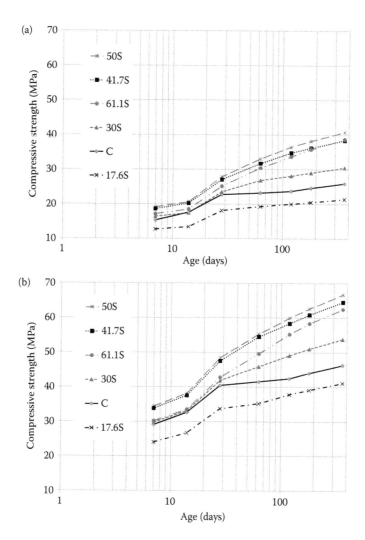

Figure 10.21 Strength development of GGBFS-incorporated concretes in comparison with the control concretes designed for a 28-day compressive strength of (a) 20 MPa and (b) 40 MPa. (Data from Öner, A., Akyüz, S. 2007. *Cement and Concrete Composites*, 29, 505–514.)

Another important feature is that the rate of strength development after 28 days is higher in GGBFS-incorporated concretes when compared to that of the control. The strength gain between 28 and 180 days was higher in all GGBFS concretes and it increases with the increasing amount of slag, as illustrated in Figure 10.22.

Studies on direct partial PC replacement by GGBFS indicate that the loss in early strength is compensated at ages at and beyond 7 days. The time required for the compensation depends on the amount of GGBFS used and its fineness. An increased amount of GGBFS results in longer, whereas increased fineness results in shorter compensation times (Babu and Kumar, 2000; LaBarca et al., 2007; Abdelkader et al., 2010).

Silica Fume: Before a discussion on the influence of SF incorporation on concrete strength, it should be noted that comparison of concretes with and without SF should be made on the basis of equal water–cementitious ratio since water reducing or high range water reducing chemical admixtures are almost always used in concretes containing more than 5% (by mass of PC) SF.

SF is a very reactive pozzolan due to its extreme fineness and almost completely amorphous siliceous character. Its effect on the concrete strength has several aspects such as pore refinement, matrix densification and pozzolanic

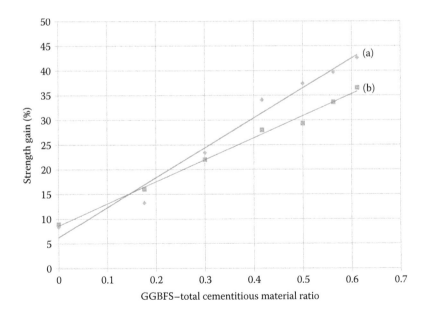

Figure 10.22 Strength gain by GGBFS incorporation between 28 and 180 days, the controls of which were designed for a 28-day compressive strength of (a) 20 MPa and (b) 40 MPa. (Data from Öner, A., Akyüz, S. 2007. *Cement and Concrete Composites,* 29, 505–514.)

reaction. However, its strengthening of the ITZ between the aggregates and the matrix is more significant than other roles it play (Siddique, 2011).

Wong and Abdul Razak (2005) studied the effect of SF on compressive strength development in three series of concretes with water–cemetitious ratios of 0.27, 0.30 and 0.33. In each series, there were four groups with 0% (control PC concrete), 5%, 10% and 15% (replacement by mass of PC) SF. The amount of high range water reducing agent was kept constant in each series as 1.8%, 0.8% and 0.5%, respectively. Their results which are given in Figure 10.23 indicate that SF did not result in strength enhancement at early ages but SF-incorporated concretes achieved higher strengths than the control concretes after 7 days. Although the early strengths were slightly lower than those of the control concretes, the strength gain of SF concretes between 1 and 7 days was 1.5–2.5 times more. On the other hand, it is necessary to note that the rate of strength gain beyond 28 days was drastically slowed down.

This was also observed by S. Wild et al. (1995) and M. Mazloom et al. (2004). Results of Wild et al. (1995) given in Figure 10.24, indicate that the relative strengths of concretes containing lower (12% and 16%) SF were reduced after 28 days, that of 20% SF-incorporated concrete remained almost the same whereas those of concretes with higher (24% and 28%) SF contents continued to increase. This was attributed to the formation of a layer of hydration products surrounding the SF particles that prevents further pozzolanic reactions. For smaller amounts of SF, the impervious layer formation was stated to be fully developed whereas for higher SF contents, it may not be sufficient to thoroughly cover all SF particles thus, pozzolanic reactions continue at later ages.

The study by Sun and Young (1993) in which the amount of unreacted SF in cement pastes containing 18%–48% additional SF was determined at various ages by 29-Si NMR supports the explanation given earlier.

Limestone Powder: Limestone powder is seldom used directly as a mineral admixture in normal concretes. Instead, it is incorporated in the concrete commonly either as an additive in portland limestone cements or as a minor additional constituent. Thus, studies on the effect of limestone powder on the strength development are generally on those of portland limestone cements or on concretes made by using portland limestone cements.

Krstulovic et al. (1994) investigated the performance of fillers in the cements which were obtained by intergrinding. The finenesses of the cements were not the same. The amount of limestone powder used was 5% and 12% (by mass of cement). The 28-day compressive strengths of limestone containing concretes were slightly less than that of the control.

Tsivilis et al. (1999) used 5%–35% limestone to obtain portland limestone cements. They interground all the clinker and limestone mixtures for the same period of time. As the amount of limestone in the mixture increased, the Blaine specific surface area of the cements also increased

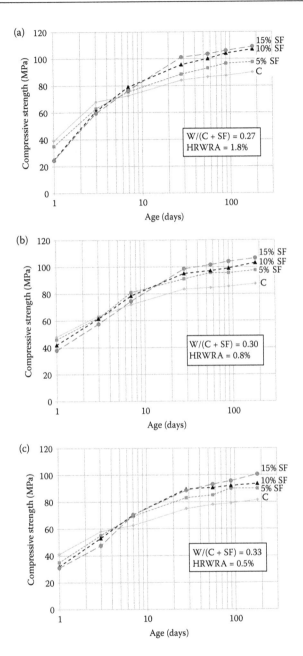

Figure 10.23 Strength development of SF-incorporated concretes with different water–cementitious ratios (a) 0.27, (b) 0.30 and (c) 0.33 in comparison with their control concretes. (Data from Wong, H.S., Abdul Razak, H. 2005. *Cement and Concrete Research*, 35, 696–702.)

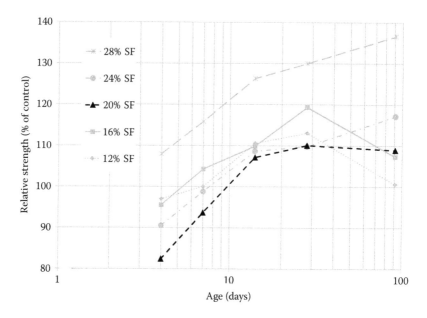

Figure 10.24 Relative strengths of SF-incorporated concretes as percent of control, at different ages. (Data from Wild, S., Sabir, B.B., Khatib, J.M. 1995. *Cement and Concrete Research*, 25(7), 1567–1580.)

due to the higher grindability of the limestone than the clinker. An example of the strength development of the standard cement mortars prepared with constant water–cementitious ratio in their study is given in Figure 10.25. It can be seen from the figure that higher or equivalent strengths may be obtained to about 15% limestone incorporation. However, a possible drawback of intergrinding would be that the remaining of the clinker portion of the blended cements are coarser. Strength data of the several cements having similar Blaine specific surface areas (369–383 m²/kg) were extracted to plot Figure 10.26, which indeed shows that at all levels of limestone incorporation there occurs strength reduction at all ages. In another investigation on the effects of intergrinding and separate grinding (Delibaş, 2012), cements prepared by separate grinding of the clinker and the limestone to the same fineness and then blending resulted in higher strengths than similar cements prepared by intergrinding, at all ages, as illustrated in Figure 10.27.

Ramezanianpour et al. (2009) have used 5%–20% (by mass) limestone powder to replace ordinary PC in three different series of concretes with water–cementitious ratios of 0.37, 0.45 and 0.55 and found that upto 10% cement replacement by limestone powder no significant strength reduction occurs at any age upto 180 days. In concretes with lower water–cementitious ratios, even slightly higher strengths than that of the control were obtained.

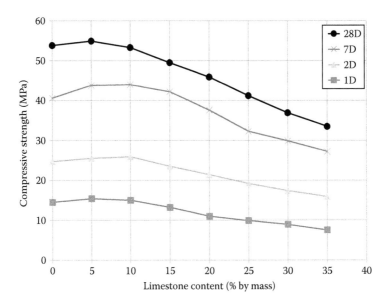

Figure 10.25 Effect of limestone content on the strength of portland limestone cements with different fineness at different ages. (Data from Tsivilis, S. et al. 1999. *Cement and Concrete Composites*, 21, 107–116.)

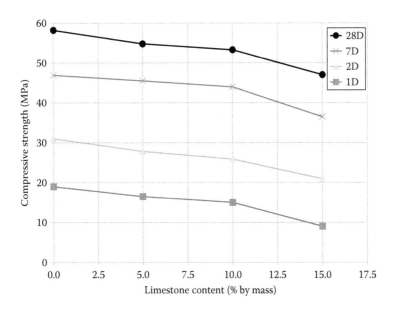

Figure 10.26 Effect of limestone content on the strength of portland limestone cements having similar fineness at different ages. (Data from Tsivilis, S. et al. 1999. *Cement and Concrete Composites*, 21, 107–116.)

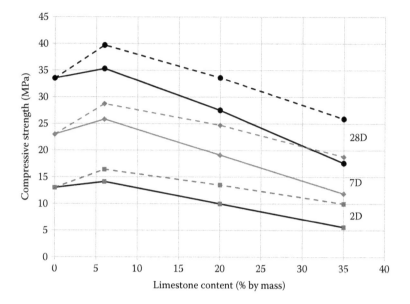

Figure 10.27 Comparison of intergrinding (full lines) and separate grinding (dashed lines) on the strength of portland limestone cements at different ages. All cements had 300 ± 10 m²/kg Blaine specific surface area. (Data from Delibaş, T. 2012. Effects of granulated blast furnace slag, trass and limestone fineness on the properties of blended cements, MS thesis, The Graduate School of Natural and Applied Sciences, METU, 72pp.)

However, with the increased amount of limestone in the mix beyond 10%, the strengths were reduced.

10.1.3.2 Effect of curing temperature

The reactivity of pozzolans and latent hydraulic mineral admixtures is influenced by the curing temperature. As all chemical reactions, hydration and pozzolanic reactions are accelerated with increasing temperature.

The effect of increased curing temperature as a means of increasing the reactivity of natural pozzolans was studied by Shi and Day (2001) on 80% natural pozzolan and 20% lime mixtures. They have measured the strength development and amount of reacted CH at curing temperatures of 23°C, 50°C and 65°C and found out that the rate of pozzolanic reaction was (1) increased within the first day with increasing temperature; (2) slowed down later at 65°C with respect to 50°C and (3) total amount of CH consumed was the highest at 50°C and the difference between those of 23° and 65°C was very small. The strength developments of their natural pozzolan–lime mixtures cured at different temperatures are shown in Figure 10.28.

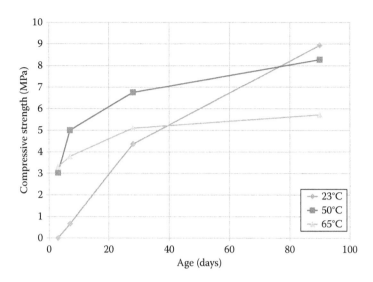

Figure 10.28 Strength development in 80% natural pozzolan and 20% lime mixtures at different curing temperatures. (Data from Shi, C., Day, R.L. 2001. *Cement and Concrete Research*, 31, 813–818.)

Another study (Ezziane et al., 2007) on the effect of different curing temperatures (20°C, 40 °C and 60°C) on the strength development of mortars containing different amounts of natural pozzolans (10%–40%, by mass) to partially replace the PC has revealed that (1) early age strengths are lower in pozzolan-incorporated mortars when compared to PC mortars and the difference increases with increasing amount of natural pozzolan and decreases with increasing curing temperature; (2) strengths of pozzolan-incorporated mortars were higher than those of PC mortars, even upto 30% replacement level, beyond 28 days at all curing temperatures studied; (3) the adverse effect of higher curing temperatures on late strengths is not as pronounced in pozzolan-incorporated mortars. The relative change in strength with respect to 20°C curing, of the PC and 40% natural pozzolan-incorporated mortars cured at 60°C is shown in Figure 10.29.

Similar behaviour also holds true for FA-incorporated concretes. However, they were found to be more sensitive to temperature effects particularly within very early ages. The strength increases in concretes with 10%, 20% and 30% (by mass) FA replacing the PC were found to be 1.6–1.9 times more than that of the control concrete within the first 10 h on increasing the curing temperature from 20°C to 40°C (Maltais and Marchand, 1997).

The effect of curing temperature on the strength development of 50:50 PC–GGBFS cement concrete compared to PC concrete is given in Figure 10.30. Increased curing temperatures are more favourable in GGBFS-incorporated concretes. However, opposite is true for low curing temperatures (Wainwright, 1986).

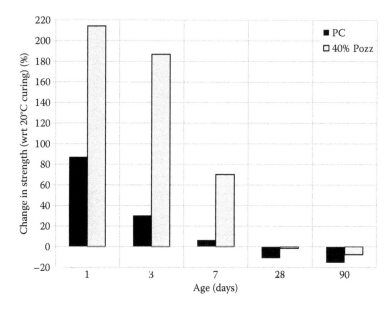

Figure 10.29 Comparison of the relative change in strength of PC and pozzolanic cement (40%, by mass natural pozzolan) at 60°C curing with respect to 20°C curing. (Data from Ezziane, K. et al. 2007. *Cement and Concrete Composites*, 29, 587–593.)

Wild et al. (1995) cured the SF-incorporated concretes at 20° and 50°C and determined that early strength is enhanced significantly by high temperature curing but at ages beyond 4 days only a minimal strength increase was observed, as illustrated in Figure 10.31. High temperature curing results in the rapid hydration of PC and formation of more calcium hydroxide at early ages. The substantial gain in early strength in SF concrete is due to the increased rate of reaction of SF with the lime produced on hydration.

10.1.3.3 Changes in matrix porosity and ITZ

Many macro properties of cementitious systems are strongly related with the microstructure. As it was already discussed earlier in this section, porosity and ITZ play a governing role in strength. Incorporation of mineral admixtures surely influence the porosity, pore size and characteristics of the ITZ.

Mercury intrusion porosimetry (MIP) tests carried out on four series of concretes as control PC concrete and GGBFS, SF and MK-incorporated (replacing 10% of PC, by mass) concretes revealed that total intrusion volume was less in mineral admixture containing concretes than the control at all ages tested between 3 and 180 days. The reduction in total porosity was attributed to the densification of the paste both by physical pore-filling

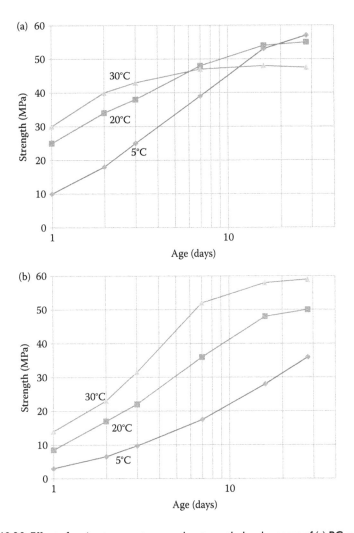

Figure 10.30 Effect of curing temperature on the strength development of (a) PC concrete and (b) 50:50 PC–GGBFS cement concrete. (Data from Wainwright, P.J. 1986. Properties of fresh and hardened concrete incorporating slag cements, in *Cement Replacement Materials* [Ed. R.N. Swamy], Surrey University Press, 100–133.)

effect and pozzolanic reactions. Furthermore, the mineral admixtures resulted in pore refinement, especially at ages beyond 28 days (Duan et al., 2013). Micro hardness determinations were carried out at various distances (10–140 μm) from the aggregate surface in the same study. It was found that all three mineral admixtures used resulted in densified ITZ which led to improved micro hardness of the ITZ, significantly. In another investigation, the influence of FA on the CH orientation at ITZ was studied and it

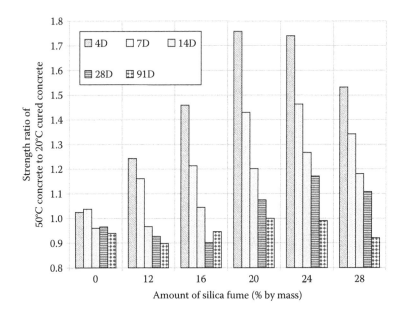

Figure 10.31 Effect of curing temperature on strength of SF concretes at different ages. (Data from Wild, S., Sabir, B.B., Khatib, J.M. 1995. *Cement and Concrete Research*, 25(7), 1567–1580.)

was found that the preferential orientation of CH crystals was much less in FA concretes than PC concrete (Jiang, 1999).

The interfacial bond strength in FA concretes was found to be lower than that of PC concrete, at early ages. Beyond 28 days, however, it becomes stronger than the PC concrete with equivalent matrix strength (Wong et al., 1999). On the other hand, the interfacial bond strength depends on the strength of the paste. The higher the paste strength, the higher becomes the bond strength (Jiang, 1999). 20% SF replacement of PC resulted in higher interfacial bond strength which was attributed to thinner and denser ITZ caused by SF (Çalışkan, 2003).

10.2 MODULUS OF ELASTICITY

The modulus of elasticity of a material is a property that describes the resistance of the material to deformation within the elastic range. 'Ideal' elasticity is exhibited by a linear stress–strain relationship which is independent of time. When the load applied on the body is removed immediate recovery occurs. The slope of the stress–strain diagram of an ideal elastic body is its Young's modulus, E. Truly elastic deformations are small and reversible.

Most engineering materials, including concrete, are not truly elastic. The stress–strain curves of such materials are curvilinear. Strain is neither directly proportional to the instantaneous stress applied nor it is fully recovered on the removal of the stress. The reason for the non-linearity of the stress–strain curve of concrete is explained by the presence of microcracks at the ITZ between the binding medium and the aggregates, even before any loading. These microcracks develop due to the differences in the stifnesses of the aggregates and cement paste that lead to differential strains. The growth of the microcracks in length, width and number under increased stress results in the non-linearity of the stress–strain curve. Typical stress–strain curves of a normal aggregate, a hardened cement paste and a concrete are shown in Figure 10.32.

Modulus of elasticity, E is affected by the characteristics of the aggregate, cement paste and ITZ. All factors affecting the density of concrete also affect E. Since density and porosity are inversely proportional, the porosities of the aggregate, the cement paste and the ITZ are the prime factors influencing the modulus of elasticity of concrete. Thus, any change in the porosity of the concrete caused by mineral admixture incorporation would affect its modulus of elasticity.

A study on four series of high strength concretes prepared by partial replacement of PC by different amounts of SF, MK, FA and GGBFS revealed that these mineral admixtures resulted in 9.7%–37.0% reduction in the porosity with respect to the control concrete. In the same study, volumes of pores <15, 15–30, 30–50 and >50 nm were also investigated and it was

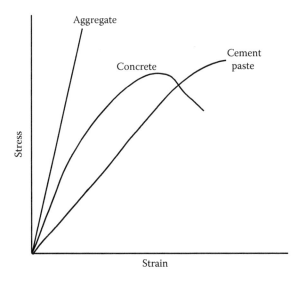

Figure 10.32 Typical stress–strain curves for normal aggregate, hardened cement paste and concrete.

found out that mineral admixtures increased the volume of pores smaller than 30 nm and decreased the volume of pores larger than 30 nm when compared with those of control PC concrete (Megat Johari et al., 2011). Their data were used to relate the modulus of elasticity to porosity in Figure 10.33. Although the correlation is rather poor, most probably due to the limitations of the MIP method used, there is a decreasing trend in modulus of elasticity with increasing porosity.

On the other hand, since both the modulus of elasticity and strength of concrete are affected by the porosity, although not to the same degree, data from the two different studies (Nassif et al., 2005; Megat Johari et al., 2011) were used to plot the strength–modulus of elasticity relationship given in Figure 10.34. Nassif et al. (2005) used FA, SF and blends of FA, SF and GGBFS in different amounts to partially replace PC.

As indicated in Figure 10.34, the relationship between strength and modulus of elasticity seems to be similar in mineral admixture-incorporated concretes and PC concretes. In other words, similar strength concretes have similar moduli of elasticity whether they contain mineral admixtures or not. Furthermore, the equations proposed to predict the modulus of elasticity from compressive strength may equally be used for mineral admixture-incorporated concretes, as illustrated in the same figure for the formula proposed by ACI (ACI 318, 2008).

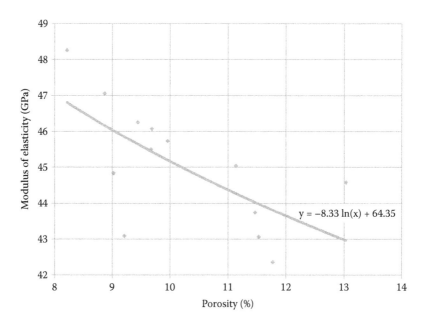

Figure 10.33 Porosity–modulus of elasticity relationship for mineral admixture-incorporated concretes. (Data from Megat Johari, M.A. et al. 2011. *Construction and Building Materials*, 25, 2639–2648.)

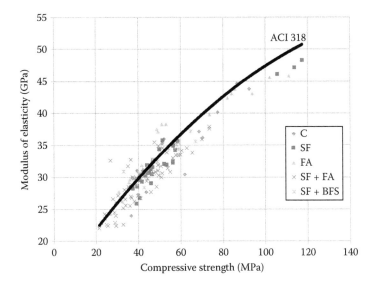

Figure 10.34 Strength–modulus of elasticity relation of blended cement concretes in comparison with PC concretes. (Data from Nassif et al. 2005; Megat Johari, M.A. et al. 2011. *Construction and Building Materials*, 25, 2639–2648.)

10.3 SHRINKAGE

Shrinkage is generally associated with the loss of moisture from concrete in either fresh or hardened state. However, such an approach to the shrinkage of concrete oversimplifies it. Shrinkage may also occur even in sealed concrete due to self-desiccation on hydration which is a phenomenon encountered in concretes with low water–cement ratios. Although different aspects of shrinkage were known since the early twentieth century, there is still a dispute on the terminology related with it. ACI defines several different forms of shrinkage as plastic, drying, carbonatation and settlement shrinkages (ACI 116, 2005). Settlement shrinkage is referred to autogeneous volume change and thermal volume change is also defined in the report.

Terms such as chemical shrinkage, self-desiccation shrinkage, autogeneous shrinkage, setting shrinkage, hydration shrinkage, etc. appear in many of the recent publications as cited by Jensen and Hansen (2001), Bentz and Jensen (2004), Šahinagić-Isović et al. (2012) and Allena and Newtson (2011) besides those defined by ACI. Definitions of the different types of shrinkages that may be encountered in cementitious systems, as given in several publications, are summarised in Table 10.2. It is not intended to make a critical review of different terms related with shrinkage here. However, it seems that an agreement on the terminology related with concrete shrinkage is necessary.

Table 10.2 Definitions of different types of concrete shrinkage

Type of shrinkage	Definition
Plastic	Shrinkage that takes place before the cementitious system sets (ACI 116, 2005). Shrinkage occuring while concrete is still fresh (Allena and Newtson, 2011).
Drying	Shrinkage resulting from loss of moisture (ACI 116, 2005). Shrinkage caused by loss of water from hardened concrete (Allena and Newtson, 2011).
Autogeneous	Shrinkage of an unrestrained, sealed cementitious system at a constant temperature (Jensen and Hansen, 2001; Bentz and Jensen, 2004). Shrinkage caused by hydration of cement (Allena and Newtson, 2011).
Settlement	Shrinkage befor the final set caused by settling of the solids and displacement of fluids (ACI 116, 2005).
Chemical	Shrinkage associated with the hydration reactions (Jensen and Hansen, 2001; Allena and Newtson, 2011).
Self-dessication	Shrinkage of a set cementitious material caused by reduction of the internal RH upon hydration reactions (Jensen and Hansen, 2001).
Setting	Shrinkage before setting (Jensen and Hansen, 2001).
Hydration	Shrinkage due to contraction of the hydration products (Šahinagić-Isović et al., 2012).
Carbonatation	Shrinkage resulting from carbonatation (ACI 116, 2005).
Thermal	Shrinkage resulting from temperature changes (Šahinagić-Isović et al., 2012).

Since some of the shrinkage types given in Table 10.2 mean almost the same thing or are either the cause or the result of each other and the consequences of shrinkage are of more practical concern than the types, only the three basic types of plastic, autogeneous and drying shrinkage will be considered here. Briefly, plastic shrinkage is caused by loss of water before the concrete hardens; autogeneous shrinkage is due to the consumption of water in the concrete by the hydration process that leads to the phenomenon known as self-dessication and drying shrinkage is caused by loss of water from hardened concrete. Autogeneous shrinkage is predominant in concretes with low water–cement ratios, whereas the other two may be observed in any concrete.

The problem associated with shrinkage is cracking due to the induced stresses that exceed the tensile strength of concrete. Usually the strains developed on plastic and autogeneous shrinkage are small, although there are several cases reporting the opposite (Paillére et al., 1989; Allena and Newtson, 2011). No matter if they are small or not, the cracks developed on plastic and autogeneous shrinkage may grow considerably with further drying and result in serious serviceability problems in concrete.

Shrinkage is a phenomenon that occurs in the binding medium (the cement paste) and aggregates in concrete have a restraining effect on shrinkage. The parameters affecting shrinkage may be grouped into three as (1) environmental parameters (RH, temperature, rate of moisture loss, duration of moisture loss); (2) geometry of the concrete element (surface area–volume ratio, thickness) and (3) cementitious paste parameters (water–cementitious ratio, amount and composition of the cementitious material, degree of hydration). Since the first two groups of parameters are not affected by the use of mineral admixtures in concrete, only the paste parameters will be dealt with here.

Drying shrinkage strains of mortars having a constant water–cementitious ratio of 0.50 and containing 0%–30% SF were measured at 28 and 405 days. Although the 28-day drying shrinkage was higher in SF-incorporated mortars and it increased with increasing SF content, very close values were recorded at 405 days (Rao, 1998). 5%–10% SF incorporation was found to reduce the drying shrinkage of high strength concretes at all ages. Triple blends containing 20% (by mass) FA or GGBFS, in addition to 10% SF, on the other hand were reported to show higher drying shrinkage than SF concretes. However, the shrinkage strains of the triple blend concretes investigated were similar to those of the reference PC concretes (Haque, 1996).

Similarly, FA incorporation was found to reduce the drying shrinkage (Chindaprasirt et al., 2004). To investigate the effect of FA fineness on drying shrinkage, two groups of FA were obtained by (a) sieving the original FA through 74 and 44 μm sieves and (b) separating the fine, medium and coarse portions of the original FA by an air separator. These classified fly ashes were used in the same amount (replacing 40% of PC, by mass) as the original FA in the mortar mix. The test results are summarised in Figure 10.35.

Shrinkage of mortars containing different mineral admixtures was also studied by Itim et al. (2011). Drying shrinkage and autogeneous shrinkage strains of limestone powder, natural pozzolan and GGBFS-incorporated mortar specimens were recorded for 1 year. Measurements started 1 day after casting the specimens. The amount of mineral admixtures replacing the PC (by mass) were 5%, 15% and 25% for limestone powder, 10%, 20% and 30% for natural pozzolan and 10%, 30% and 50% for GGBFS. Water–cementitious ratio was 0.47 for all mixtures. Although the autogeneous shrinkage within the first 3 days increased with increasing limestone powder content, later age values were found to be similar to those of the control mortar. Similarly, drying shrinkage strains were comparable with those of the control mortar at all ages and for all limestone contents. A similar behaviour was also observed in natural pozzolan-incorporated mortars. Autogeneous shrinkage in slag mortars was slightly higher at 10% replacement level at all ages, except 3 days. However, for higher slag contents, it was reduced considerably. Drying shrinkage, on the other hand, decreased at all ages for 10% slag content and increased with increasing

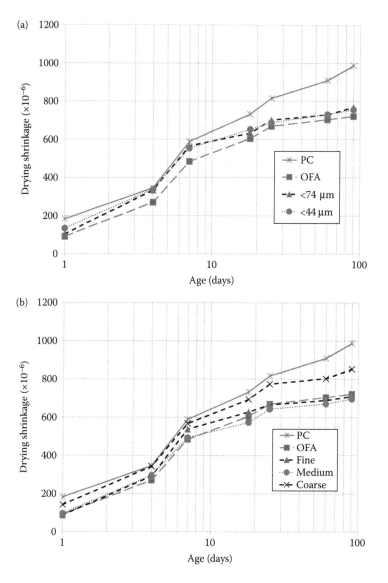

Figure 10.35 Effect of FA fineness on drying shrinkage of FA mortars. Original FA was (a) sieved or (b) air-separated. (Data from Chindaprasirt, P., Homwuttiwong, S., Sirivivatnanon, V. 2004. *Cement and Concrete Resarch*, 34, 1087–1092.)

slag amount (Itim et al., 2011). On the other hand, another study on concretes containing 50% and 70% (by mass) GGBFS and with 0.30, 0.42 and 0.55 water–cementitious ratios in comparison with reference PC concretes with the same water–cement ratios revealed that total shrinkage strains were similar, at all ages (Dellinghausen et al., 2012).

The effect of GGBFS and additional SF on drying shrinkage of high strength concretes was studied by replacing (a) 30% of PC by GGBFS, (b) 40% of PC by 30% GGBFS and 10% SF. Shrinkage measurements started after 28 days of curing and continued for 180 days. The strains were lower than those of the control at all ages and use of additional SF resulted in lower drying shrinkage than in the concrete with GGBFS (Li and Yao, 2001).

In general, proper mix proportioning of mineral admixture-incorporated concrete with due care on water–cementitious ratio and aggregate content, results in similar, slightly lower or higher drying shrinkage than that of an equivalent PC concrete.

Autogenous shrinkage is not as significant in normal concretes as it is in high-strength concretes. It may be as high as drying shrinkage in low water–cementitious ratio concretes. Although many different methods used for measuring and numerous different approaches taken for the interpretation of the results (Craeye et al., 2010; Eppers, 2010) make it very difficult to discuss the effects of mineral admixtures on autogeneous shrinkage, some general remarks can be made: (1) Lower water–cementitious ratios lead to higher autogenous shrinkage (Jiang et al., 2014) which is also true for concretes without any mineral admixtures (Tazawa and Miyazawa, 1995). (2) There occurs an autogenous expansion within the first 24 h which is attributed to ettringite (Termkhajornkit et al., 2005) and/or calcium hydroxide formation (Baroghel-Bouny et al., 2006) or disjoining pressure caused by the water adsorption on the surface of the mineral admixture particles (Craeye et al., 2010). The expansion is followed by contraction due to capillary tension. (3) As long as the pore structure and average pore diameter of the pastes are similar, autogenous shrinkage does not change significantly whether the paste contains mineral admixtures or not (Craeye et al., 2010). In other words, refined pore structure by mineral admixture incorporation may lead to higher autogenous shrinkage (Li et al., 2010). (4) As the fineness of the mineral admixture is increased, autogenous shrinkage is also increased (Craeye et al., 2010).

10.4 CREEP

Creep can be defined as the progressive deformation under a sustained load. The most general form of the strain–time curve of a material under creep is shown in Figure 10.36. The curve consists of three parts as (a) primary (transient) creep which is similar in nature to delayed elastic deformation; (b) secondary (steady-state) creep and (c) tertiary (accelerated) creep which are viscous in character. Primary and secondary creep stages cannot easily be distinguished from each other and tertiary creep stage does not exist in concretes subjected to normal sustained stress levels (Neville et al., 1983), as illustrated in Figure 10.37.

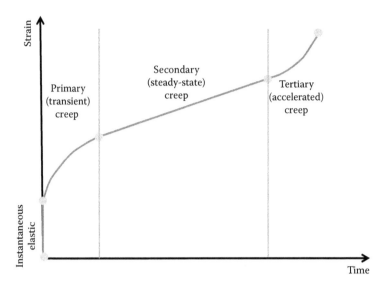

Figure 10.36 A typical strain–time curve at constant stress and temperature.

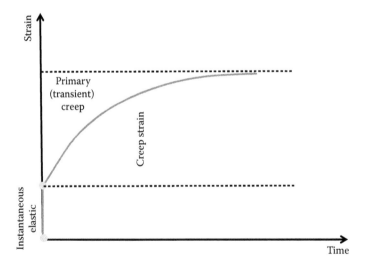

Figure 10.37 A typical strain–time curve for concrete under normal sustained stress levels.

On the removal of the stress concrete undergoes an instantaneous elastic recovery which is approximately the same as the instantaneous elastic strain. It is followed by a time-dependent recovery which occurs faster than the creep. However, the recoverable strain is less than the creep and there remains a residual strain.

Creep of concrete is dependent on numerous factors such as (1) type, composition and fineness of cement; (2) type of chemical admixture used; (3) type, amount, maximum size and modulus of elasticity of aggregate; (4) water–cement ratio; (5) strength and magnitude of stress applied; (6) age; (7) shape and size of the concrete member and (8) ambient humidity and temperature.

Creep and shrinkage of concrete are often considered as interrelated phenomena due to various similarities between them: Their strain–time curves, the magnitude of strains, most of the factors affecting them, both being paste properties and aggregates acting as restraints and both having considerable irreversibility. However, their mechanisms are different (Mindess and Young, 1981). Generally, both phenomena occur simultaneously and the common practice considers that they are additive for compressive stresses, which is practical but not totally correct. In fact, it is known that shrinkage increases the creep strain as illustrated in Figure 10.38 (Mindess and Young, 1981; Neville et al., 1983).

The amount of creep strain of a given concrete with a specified time of loading depends on the ratio of the stress applied to the strength. Therefore, two useful concepts named as creep coefficient and specific creep are commonly utilised to compare the creep behaviour of different concretes. Creep coefficient (C) is the ratio of the creep strain to the instantaneous elastic strain whereas specific creep (ø) is the ratio of creep strain to the stress applied. Since mineral admixture incorporation in concrete affects both the strength and strength development rate, comparison of concretes with and without mineral admixtures with respect to their creep behaviour would be more meaningful when these two concepts are made use of.

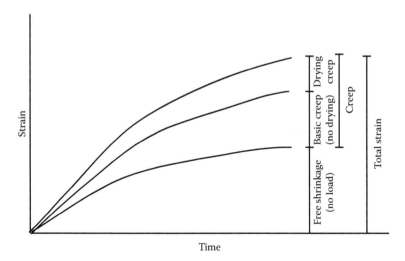

Figure 10.38 Creep of concrete under simultaneous drying and sustained load. Note that the curves are in excess of the instantaneous elastic strain.

The literature contains numerous reports on the effect of mineral admixtures on the creep of concrete. Owing to the considerable differences of the independent variables such as age of loading, stress–strength ratio, amount of aggregate and type of concrete, the data available in the reports are far from making general conclusive remarks. Nevertheless, results from several researches are summarised.

Creep tests on three concretes (A, B and C) with 28-day compressive strengths of 81.1, 100.4 and 104.0 MPa were loaded to 40% of their strength at 28 days and loading continued for 180 days. Concrete A was the control whereas concrete B had 30% GGBFS and concrete C had 30% GGBFS and 10% SF replacing the PC, by mass. All other mixture parameters were held constant. Both the creep strain and its rate were lower in mineral admixture incorporated concretes than those in the control. At the end of the tests, concretes A, B and C had creep values of 1293×10^{-6}, 623×10^{-6} and 450×10^{-6}, respectively (Li and Yao, 2001).

Seven different concretes with 430 kg/m^3 of cementitious material and 0.35 water–cementitious ratio were used to study the specific creep by the method described in ASTM 512 (Standard Test Method for Creep of Concrete in Compression). The control concrete made by using PC was compared with concretes made by using cements with (a) 65% GGBFS, (b) 10% SF, (c) 10% SF and 15% FA and (d) 10% SF and 25% FA, replacing the PC, by mass. Besides these, two other concretes were prepared by using 10% (by mass) SF to replace (e) the cement in (a) and (f) a slag cement with 35% GGBFS. Specific creep was determined to be higher as the amount of slag in the cementitious material increases. On the other hand, SF incorporation reduced the specific creep considerably with respect to PC and marginally with respect to the slag cements. Additional FA resulted in specific creep values much lower than the PC but more than SF incorporation (Khatri et al., 1995).

In a study, on the time-dependent deformations of limestone powder, three types of SCCs (SCC1, SCC2 and SCC3) are taken here to discuss the effect of limestone powder content. Total cementitious material content and water–cementitious ratio were 600 kg/m^3 and 0.28 in all mixes, respectively. They had very similar paste volumes and close rheological properties. A PC (CEM I 42.5 R) was used in SCC1 and SCC3. Limestone powder-to-total cementitious material ratios were 0.4 and 0.5, respectively. A slag cement (CEM III/A 42.5 N) was used in SCC2 with a limestone powder-to-total cementitious material ratio of 0.4. Creep coefficients of SCC1 and SCC3 were similar to each other whereas SCC2 had lower values, at all ages (Heirman et al., 2008).

Both compressive and tensile specific creep of PC and SF- and GGBFS-incorporated concretes were studied on six groups of specimens: (1) PC concrete with 360 kg/m^3 cement and 0.50 water–cement ratio; (2) PC concrete with 400 kg/m^3 cement and 0.50 water–cement ratio; (3) PC concrete with 450 kg/m^3 PC and water–cement ratio of 0.30; (4) SF-incorporated concrete with 450 kg/m^3 PC, 50 kg/m^3 of SF and water–cementitious ratio of 0.30; (5) GGBFS-incorporated concrete with 175 kg/m^3 PC, 325 kg/m^3 of GGBFS and water–cementitious ratio of 0.30 and (6) SF- and GGBFS-incorporated concrete with 175 kg/m^3 of PC, 50 kg/m^3 of SF, 275 kg/m^3 of GGBFS and water–cementitious ratio of 0.30. The loading was started after 3 days of curing and the stress applied was 30% for compressive creep and 20% for tensile creep of the average 3-day respective strengths of each concrete. The main findings of the study showed that (1) tensile specific creep of all concretes was greater than the compressive specific creep; (2) high-water–cement ratio resulted in higher specific creep and (3) the blended cements had similar specific creep values and they were lower than those of the control concrete with the same water–cementitious ratio (Li et al., 2002).

It is convenient to conclude the discussion on the effect of mineral admixtures on creep of concrete by referring to the statements made in the three ACI committee reports:

1. Although published data on creep of concrete containing slag cement indicate somewhat conflicting results which are likely to be affected by differences in maturity and characteristics of the PC used with the slag, concretes containing slag cement generally have lower basic creep and similar or lower total creep compared with concrete containing only PC (ACI 233R-03, 2003).
2. The effect of FA on creep of concrete primarily depends on the extent to which it influences the strength and rate of strength development. FA concretes proportioned to have the same strength at the age of loading produce less creep strains than the equivalent PC concretes (ACI 232.2R-03, 2003).
3. It is difficult to draw specific conclusions on the effect of SF on the creep of concrete due to the limited published data and differences in creep tests used by various researchers. However, it can be stated with certainty that the creep of SF concrete is not higher than the creep of concrete of equal strength without SF (ACI 234R-96, 2000).

Chapter 11

Effects of mineral admixtures on durability of concrete

Durability of a material may be defined as the ability to exist for a long time without significant deterioration, by resisting the environmental effects which result in some changes in the properties of the material with time. Service life and durability are usually considered as equivalent terms. Under given conditions of use, if the properties deteriorate to such an extent that safety or economy outrules further use, the material is considered to reach the end of its service life (Mehta and Monteiro, 2006). Durability is not only significant from the safety, economy and technical points of view. Ecologically, materials with longer service lives mean better conservation of natural resources.

The durability of concrete is defined as its ability to withstand weathering action, chemical attack, abrasion or any other process of deterioration (ACI 201.2R-01, 2001). Although it is a durable material as long as it is properly designed and carefully produced, there are several potentially harmful processes that concrete may encounter. The common processes are described below. It should be noted that these processes may be of chemical or physical nature or both; when more than one process is encountered simultaneously or one after the other, their harmful effects may be aggravated; and most of them involve water either as the deteriorating agent itself or providing a means for the destructive process.

11.1 FREEZING AND THAWING

There occurs a volume expansion slightly over 9% when water turns into ice upon freezing. Like any other porous material, saturated or nearly saturated concrete may suffer from freezing and thawing unless it is designed properly to resist it. Unless it is more than 91% saturated, the empty space will be able to take in the expansion without causing any disruptive internal stress. However, the whole concrete body need not be saturated in order to be affected by freezing. The portions close to the surface may be damaged even though the inner portions may remain intact (Mays, 1992).

Hydrated cement paste contains different types of voids as gel pores, capillary voids and air voids. Gel pores have sizes ranging from 0.5 to 2.5 nm. The sizes of capillary voids depend on the water–cement ratio and degree of hydration. Although the usual size range in a well hydrated cement paste with low water–cement ratio is 10–50 nm, some capillary voids may be as large as 5 μm. Entrapped air voids are usually between 1–3 mm in size whereas the entrained air voids are 50–200 μm (Mehta and Monteiro, 2006). The water in the largest voids is the first to freeze. As the pore size gets smaller lower temperatures are necessary for the water to freeze. For example, water in the pores of 10 nm size freezes at –5°C, whereas water in the pores of 3.5 nm size freeze at –20°C (Mindess and Young, 1981). Water adsorbed by or held in the gel pores remains unfrozen at temperatures above ~–78°C (Mehta and Monteiro, 2006).

11.1.1 The frost attack mechanism

The deterioration of concrete upon freezing and thawing cannot be explained by a single mechanism. It seems that different mechanisms may act together or sometimes may counteract. The three theories that try to explain this damaging process are summarised below.

11.1.1.1 Hydraulic pressure theory

As ice starts to form in the capillaries, the volume increase forces the remaining unfrozen water to be compressed. The hydraulic pressure thus created may be relieved if the residual water can diffuse from the capillary to a free space. In a non-air-entrained concrete the free space is usually the outer surface which may either be impossible or takes too long to access. Therefore, the capillary tries to expand and applies stress to the paste around. When the total stress from the adjacent capillaries reaches the tensile strength of the paste, cracking occurs.

11.1.1.2 Osmotic pressure theory

The water in the capillaries contains various dissolved salts such as alkali sulphates and chlorides. Thermodynamically, the free energy of water in the solution is less than that of pure water. If there occurs a salt concentration gradient it results in the movement of water from higher energy state to the lower one (Rønning, 2001). When ice starts to form in a large capillary, the salts in the solution would concentrate in the non-frozen portion of the solution close to the capillary walls. The increased salt concentration results in a gradient that facilitates the movement of water from the surroundings to the ice containing capillary, ending up with more and more ice formation and dilation of the capillary.

The two theories given above were proposed by Powers (1958) and his co-workers. They are stated to be acting simultaneously. However, there is not much evidence that the osmotic pressure arises in the bulk of the concrete although it may be observed at the zones close to the concrete surfaces where deicing salts are applied (Rønning, 2001).

11.1.1.3 Litvan's theory

While the water in the large voids starts to freeze the water that is in the gel pores remains as a supercooled liquid, which is in a higher energy state than the ice formed. The difference in their energy states forces the water in the gel pores to move to places with lower energy state (larger voids) which results in more ice formation and dilation. Upon thawing, further internal pressure caused by the water increases the crack formation (Litvan, 1976).

Although the disruptive effect of freezing–thawing is mainly observed in the cement paste, certain aggregates may be sensitive to frost action, also. Generally, aggregates with very low and very high porosity do not pose a durability problem since they either do not saturate readily in concrete or the water can easily escape during freezing. However, fine-grained aggregates such as sandstone with fine pores and high total porosity may be critical.

11.1.2 Influence of air-entrainment on freeze–thaw resistance

Evidently, whatever the freezing mechanism involved, the internal stresses developed should be relieved in order to prevent damage. Air-entrainment which is simply the creation of very small air bubbles in the cement paste is a standard practice to improve the freezing–thawing resistance of concrete. The idea behind air-entrainment is to provide empty spaces to which the residual water upon the start of freezing would easily migrate and the stresses induced are relieved. Usually, 4%–8% air-entrainment (by volume of concrete) is recommended. The air content varies with the severity of exposure and the maximum aggregate size.

Air-entraining admixtures are surfactants obtained by the processing of wood resins, by-products of paper and petroleum industries. They are characterised by hydrocarbon chain molecules with one end having hydrophilic polar group, usually a carboxylic or sulphonic acid. The molecules concentrate at the water–air interfaces and reduce the surface tension of water which facilitates air bubble formation upon mixing. Then the bubbles are stabilised by the orientation of the molecules with their hydrophilic ends in water and hydrophobic ends in the air bubbles (Edmeades and Hewlett, 1988). After setting, the air bubbles remain in the rigid mass thus forming an entrained air-void system within the concrete. Although the total entrained-air content is an important parameter, the spacing factor (the distance between the air voids) and the specific surface (number and size of air voids for a specific total volume of air) of the air-void system are also significant. The air voids in air-entrained concrete have sizes ranging from 5 μm to 1 mm. They are uniformly distributed throughout the concrete body. The spacing factor, which is the indicator of the average distance the water should travel to enter an air void and relieve the stress, should be small. Higher specific surface means more air voids with smaller dimensions. Generally, spacing factors less than 0.2 mm and specific surfaces greater than 24 mm²/mm³ are stated to result in higher freeze–thaw durability (PCA, 1998).

11.1.3 Influence of mineral admixtures on freeze–thaw resistance

The ability of a concrete to resist freezing and thawing depends basically on the adequacy of the air-entrainment. Other factors that should be considered include (a) using sound, frost-resistant aggregates and (b) ensuring sufficient strength which is usually taken as >3.5 MPa, before freezing starts, and >28 MPa for repeated freezing–thawing (Thomas, 2013). As long as these requirements are fulfilled, the risk of damage caused by freezing and thawing is greatly reduced in any concrete whether it contains mineral admixtures or not.

The effect of mineral admixtures on the entrained-air content was already discussed in Section 9.4. Basically, an increased amount of fine materials in concrete require higher dosages of AEA to achieve a specified amount of air-entrainment.

There are some conflicting data on the freeze–thaw resistance of mineral admixture-incorporated concretes in the literature. It seems that generally comparisons are not systematically made on concretes having similar air-void systems and/or similar strength. Therefore, in most reviews on the effect of mineral admixtures on the durability of concrete, the freeze–thaw resistance of mineral admixture-incorporated concretes is stated to be not different from that of PC concretes of similar strength and air-void system (Massazza, 1988; Hawkins et al., 2003; Thomas, 2013).

11.1.4 Influence of mineral admixtures on deicer salt scaling

A major durability issue of concrete pavements and walkways encountered in cold climates is the deicer salt scaling. It is a damage caused by saline solution on the concrete surface. It may be thought as not affecting the overall integrity of the concrete other than being unsightly. However, it is a progressive process and once the chips or flakes of concrete from the surface are removed, further ingress of aggressive liquids and gases deeper into concrete becomes easier (Valenza II and Scherer, 2007a).

Several mechanisms were proposed for deicer salt scaling. However, none of them were able to explain the process thoroughly. Recently, another mechanism was proposed by creating a similitude to the glue spalling technique used to decorate glass surfaces (Valenza II and Scherer, 2007b): When the salt solution on the concrete surface freezes it tries to contract much more than the underlying concrete and cracks. These cracks penetrate the concrete surface, propagate parallel to the ice–concrete interface and result in spalling.

There are several points that should be noted about deicer salt scaling: (1) it is not correlated with the internal frost resistance of concrete, (2) air-entrainment was found to be beneficial in reducing it, (3) excessive bleeding of fresh concrete results in reduced surface strength of the concrete which, in turn, reduces the deicer salt scaling resistance and (4) solutions with salt concentrations around 3% are the most destructive (Valenza II and Scherer, 2007).

Most of the studies carried out so far on the deicer salt scaling of mineral admixture-incorporated concretes were according to ASTM C 672 (1998). Almost all of them concluded that mineral admixtures increase the susceptibility of concrete for deicer salt scaling (Pigeon et al., 1996; Bouzoubaâ et al., 2001; Chidiac and Panesar, 2008; Valenza II and Scherer, 2007; Nowak-Michta, 2013; Van den Heede et al., 2013) except a few (Naik et al., 1995; Deja, 2003; Hassan et al., 2012). On the other hand, Thomas (2013) carried out extensive studies on deicer salt scaling of mineral admixture-incorporated concretes both in the laboratory and field and stated that the laboratory tests are overly aggressive and do not correlate well with field performance. He reasonably concluded that mineral admixture-incorporated concretes show good deicer salt scaling resistance as long as they are appropriately proportioned and manufactured with proper placing, compaction, finishing and curing practices.

11.2 SULPHATE ATTACK

Sulphates of calcium, sodium, potassium and magnesium in solution may enter the concrete and attack the cementitious materials resulting in expansion, cracking, spalling or softening and disintegration. Naturally occurring sulphates may be present in soils as different minerals and in dissolved form in ground waters. Concentration of sulphates in soils may vary between two points even within a short distance of each other. It may also vary with depth, depending on the climatic conditions. In hot and arid climates, sulphate concentration may be high at or close to the surface of the soil but in wet climates where the rainfall is more than evaporation, sulphates are leached down (Eglinton, 1988). Some of the common sulphate minerals found in soils are given in Table 11.1.

Sulphate concentration in ground waters may vary depending on the sulphate concentration of the soil, the movement of water, temperature

Table 11.1 Various common sulphate minerals in soils

Mineral name	General chemical formula
Anhydrite	$CaSO_4$
Hemihydrate (or bassanite)	$CaSO_4 \cdot 1/2H_2O$
Gypsum	$CaSO_4 \cdot 2H_2O$
Kieserite	$MgSO_4 \cdot H_2O$
Epsomite	$MgSO_4 \cdot 7H_2O$
Thenardite	Na_2SO_4
Mirabilite	$Na_2SO_4 \cdot 10H_2O$
Arcanite	K_2SO_4
Glauberite	$Na_2Ca(SO_4)_2$
Langbeinite	$K_2Mg(SO_4)_3$

Source: Adapted from ACI 201. 2001. Guide to durable concrete, ACI
Committee 201 Report, ACI 201.2R01, American Concrete Institute,
Farmington Hills, MI.

and the solubility of the sulphate minerals. Solubility of the sulphates is significant since it determines the amount of sulphate ions released that would react with the hydration products of cement and lead to deterioration. For example, $CaSO_4$ which has a low solubility (0.255 g/100 g at 20°C) in water would result in 0.15 g/100 g SO_3 whereas $MgSO_4$ which has a solubility of 35.1 g/100 g at 20°C would cause about 23 g/100 g SO_3 concentration in the water.

The ACI and European Concrete Standard (EN 206-1) both specify four potential sulphate exposure classes that the concrete should be protected against. The classes and corresponding sulphate concentrations are listed in Table 11.2. Exposure classes 0 in ACI 201 and XA 0 in EN 206-1 do not require any special precaution against sulphate resistance.

Table 11.2 Sulfate exposure classes specified in ACI 201 (2001) and EN 206-1 (2002)

ACI 201 (2001)			EN 206-1 (2002)		
	SO_4 concentration for			SO_4 concentration for	
Exposure class	Soils (%)	Ground waters (mg/L)	Exposure class	Soils (%)	Ground waters (mg/L)
Class 0	<0.10	0–150	XA 0	<0.20	<200
Class 1	>0.10 and <0.20	>150 and <1,500	XA 1	0.20–0.30	200–600
Class 2	0.20 to <2.00	1,500 to <10,000	XA 2	>0.30 and ≤1.20	>600 and ≤3,000
Class 3	≥2.00	≥10,000	XA 3	>1.20 and ≤2.40	>3,000 and ≤6,000

11.2.1 Action of sulphates on portland cement concrete

All soluble sulphates may lead to deleterious effects on concrete but some may be more dangerous. The severity of the attack depends also on the base of the sulphate involved. Sodium, potassium and magnesium sulphates, being much more soluble than calcium sulphates, would result in higher sulphate ion concentrations in water.

Sulphate attack on PC concrete generally involves three consecutive steps: (1) penetration of sulphate ions into concrete, (2) reaction of sulphate ions with calcium hydroxide to form gypsum and (3) reaction of gypsum with calcium monosulphoaluminate hydrate to result in ettringite. Sulphate and ettringite formation in PC paste attacked by Na_2SO_4 solution is shown in the SEM images given in Figure 11.1.

The deterioration mechanism of calcium monosulphoaluminate hydrate by magnesium sulphate is different from the others. The basic chemical equations of sulphate attack are given below:

$$Step\ 2:\quad Ca(OH)_2 + Na_2SO_4 \cdot 10H_2O \rightarrow CaSO_4 \cdot 2H_2O + 2NaOH$$
$$+ 8H_2O$$

$$(11.1)$$

or with cement chemistry abbreviations,

$$CH + N\bar{S}H_{10} \rightarrow C\bar{S}H_2 + NH + 8H \qquad (11.2)$$

$$Step\ 3:\quad 4CaO \cdot Al_2O_3 \cdot SO_3 \cdot 12H_2O + 2(CaSO_4 \cdot 2H_2O) + 16H_2O$$
$$\rightarrow 6CaO \cdot Al_2O_3 \cdot 3SO_3 \cdot 32H_2O$$

$$(11.3)$$

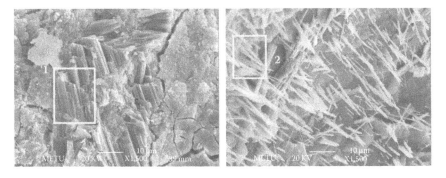

Figure 11.1 SEM images of ordinary PC paste attacked by sodium sulphate solution. Section 1 is gypsum and Section 2 is ettringite. (From Şahmaran, M., Erdem, T.K., Yaman, I.O. 2007. *Construction and Building Materials*, 21, 1771–1778. With permission.)

or with cement chemistry abbreviations,

$$C_4A\bar{S}H_{12} + 2C\bar{S}H_2 + 16H \rightarrow C_6A\bar{S}_3H_{32} \tag{11.4}$$

Note that calcium sulphate reacts only with monosulphate form and potassium sulphate behaves like sodium sulphate. These reactions result in disruptive expansion since the volumes required by the transformation of calcium hydroxide into gypsum and calcium monosulphoaluminate hydrate into ettringite are more than twice those of the attacked phases.

The action of magnesium sulphate on calcium hydroxide and calcium monosulphoaluminate hydrate results in the formation of gypsum and magnesium hydroxide as given by Equations 11.5 and 11.6. The latter reaction also yields hydrated alumina.

$$CH + M\bar{S}H_7 \rightarrow C\bar{S}H_2 + MH + 5H \tag{11.5}$$

$$C_4A\bar{S}H_{12} + 3M\bar{S} + 2H \rightarrow 4C\bar{S}H_2 + 3MH + AH_3 \tag{11.6}$$

Magnesium sulphate may attack and decompose C–S–H gel in addition to the reactions it makes with calcium hydroxide and calcium monosulphoaluminate hydrate. The reaction is given by Equation 11.7.

$$C_3S_2H_8 + 3M\bar{S}H_7 \rightarrow 3C\bar{S}H_2 + 3MH_2 + 2SH_x \tag{11.7}$$

Magnesium hydroxide and the silica gel obtained upon this reaction may further react very slowly with each other to end up with a magnesium silicate hydrate which does not have a binding value. The reason why the reaction given by Equation 11.7 occurs with magnesium sulphate, but not with other alkali sulphates is attributed to the very low solubility of the magnesium hydroxide which is around 9.6×10^{-3} g/100 g as compared to 109 and 112 g/100 g for sodium hydroxide and potassium hydroxide, respectively. The pH of the pore solution saturated with magnesium hydroxide becomes about 10.5, which is lower than the value (12.5–13.5) that is required to keep the C–S–H gel stable. Calcium silicates may liberate lime into the solution to reestablish the equilibrium, however, it is further attacked by the magnesium sulphate to produce magnesium hydroxide as given by Equation 11.5 (Eglinton, 1988). Thomas (2013) stated that alkali sulphates may also attack C–S–H gel upon prolonged exposure in a similar manner as magnesium sulphate. However, highly soluble alkali hydroxides produced prevent further attack by increasing the pH of the pore solution.

The attack of sulphates on the C–S–H gel results in softening of the cement paste with a loss of binding property. Whether expansion is accompanied

Table 11.3 Relative rate of expansion mortars prepared by different cement compound mixtures

Compound	Time to expand 0.5% in solutions with equivalent SO_3 contents of 1.2%		
	1.8% $MgSO_4$ solution	Saturated $CaSO_4$ solution	2.1% Na_2SO_4 solution
C_2S	28 days	Negligible after 18 years	Negligible after 18 years
C_3S	35 days	0.22% after 9 years	12 years
50%C_3S + 50%C_2S	65 days	0.19% expansion after 18 years	0.04% expansion after 12 years
80%C_2S + 20%C_3A	6 days	10 days	4 days
80%C_3S + 20%C_3A	4 days	11 days	7 days
80%C_3S + 20%C_4AF	16 days	0.15% expansion after 3 years	400 days
40%C_3S + 40%C_2S + 20%C_4AF	43 days	0.06%–0.07% expansion after 3 years	0.06%–0.07% expansion after 3 years

Source: Reprinted from *Lea's Chemistry of Cement and Concrete*, 4th Ed., Eglinton, M., Resistance of concrete to destructive agencies, (Ed. P.C. Hewlett), Elsevier, Oxford, Copyright 1988, with permission from Elsevier.

with this attack still remains to be a matter of debate although the researchers who support the expansion are more than those who do not (Tian and Cohen, 2000).

The relative rates of expansion of hydrated pure calcium silicates and various mixtures of calcium silicates and aluminates in magnesium, sodium and calcium sulphate solutions are given in Table 11.3, in terms of time to reach 0.5% expansion (Eglinton, 1988). The specimens used were mortar bars moist cured for 8 weeks prior to immersion into sulphate solutions. The data given in the table indicate that generalisations about the sulphate attack are not easily possible. As it is stated in most reviews, although there had been many research efforts to describe the sulphate attack since early 1900s, due to the complicated nature of the process, many of them still remain conflicting (Hyme and Mather, 1999; Santhanam et al., 2001).

11.2.1.1 Thaumasite formation

Thaumasite formation in cement paste is considered as another form of sulphate attack although it also involves carbonates. It is formed by reactions that involve calcium, silicate, sulphate and carbonate ions and sufficient water. After reviewing many field cases and laboratory investigations, Hobbs (2003) stated several conditions for thaumasite formation: (a) a low temperature (generally, below 10°C), (b) wet conditions, (c) exposure to sulphates (particularly magnesium) and/or exposure to sulphuric acid,

(d) pre-existing ettringite, (e) deficiency of calcium hydroxide and (f) inclusion of more than 10% (by mass of PC) calcium carbonate. Excessive thaumasite formation results in a soft non-cohesive mass.

The general chemical equations describing the thaumasite formation in cementitious systems are given below (Bensted, 1999):

$$\underbrace{Ca_3Si_2O_7 \cdot 3H_2O}_{\text{C–S–H gel}} + \underbrace{2(CaSO_4 \cdot 2H_2O)}_{\text{gypsum}} + \underbrace{CaCO_3}_{\text{limestone}} + CO_2$$
$$+ 23H_2O \rightarrow \underbrace{Ca_6[Si(OH)_6]_2(CO_3)_2(SO_4)_2 \cdot 24H_2O}_{\text{thaumasite}} \tag{11.8}$$

$$\underbrace{Ca_6[Al(OH)_6]_2(SO_4)_3 \cdot 26H_2O}_{\text{ettringite}} + \underbrace{Ca_3Si_2O_7 \cdot 3H_2O}_{\text{C–S–H gel}}$$
$$+ CaCO_3 + CO_2 + xH_2O \rightarrow \underbrace{Ca_6[Si(OH_6)]_2(CO_3)_2(SO_4)_2 \cdot 24H_2O}_{\text{thaumasite}}$$
$$+ CaSO_4 \cdot 2H_2O + Al_2O_3 \cdot xH_2O + 3Ca(OH)_2$$
$$\tag{11.9}$$

11.2.1.2 Delayed ettringite formation

Delayed ettringite formation (DEF) is a sulphate attack which involves internal sulphates in concrete, only. The process starts after the concrete has considerably hardened. It is generally accepted that DEF occurs in concretes that have experienced temperatures exceeding 70°C at early stages of hydration which is usually encountered in high-temperature steam curing and sometimes in mass concretes (Taylor et al., 2001). However, there had been cases reported with DEF in concretes that did not experience high temperatures (Diamond, 1996; Tosun, 2007). Furthermore, there are researchers that believe that DEF may also occur by the penetration of external sulphates (Collepardi, 2003).

The chemistry involved in DEF may be summarised as follows (Taylor et al., 2001; Tosun, 2007): at temperatures above 70°C, ettringite becomes unstable and decomposes into monosulphoaluminate hydrate and the released alumina is bound by the C–S–H gel and some hydrogarnet is formed and some of the sulphate is loosely bound by the C–S–H gel while some remains in the pore solution. After the concrete is brought to ordinary temperatures, the dissolution of C–S–H and monosulphoaluminate hydrate result in the release of some calcium, sulphate, hydroxyl and hydroxyaluminate ions that react with each other and with water to reform ettringite as given below.

$$6Ca^{2+} + 2Al(OH)_4^- + 4OH^- + 3SO_4^{2-} + 26H_2O$$
$$\rightarrow Ca_6[Al(OH)_6]_2 \cdot (SO_4)_2 \cdot 26H_2O \tag{11.10}$$

DEF is an expansive process that may be manifested as early as several weeks or as late as years in concretes that are frequently or continuously exposed to moisture (Tosun, 2006). Prior microcracking due to processes such as alkali–silica reaction or freezing–thawing may both accelerate DEF and increase the damage caused by it (Ekolu et al., 2007).

11.2.2 Influence of mineral admixtures on sulphate attack

Protection against sulphate attack is obtained by using a concrete that reduces the ingress of water and its movement within the body, which means a concrete with low permeability. A low-permeability concrete is obtained by using a low water–cement ratio. Obviously, proper placement, compaction, finishing and curing methods should also be applied. (ACI 201, 2001). For a given set of conditions and porosity, the sulphate resistance of PC concrete is affected by the compound composition of the cement used. The C_3A content is the most important parameter affecting sulphate resistance but there are evidences that the ferrite phase and the C_3S–C_2S ratio may also be effective (Tokyay and Dilek, 2003). The effect of the C_3A content of PCs on the expansion of mortar specimens is illustrated in Figure 11.2. The method described in ASTM C 1012 (1995) was used. The specimens were immersed into 5% Na_2SO_4 solution after their companion cube specimens

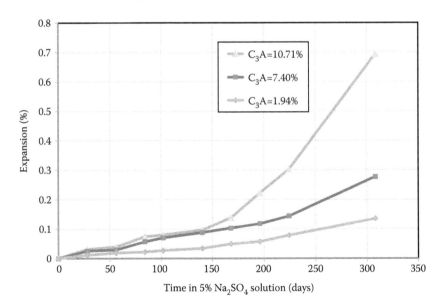

Figure 11.2 Effect of C_3A content of PC on the sulphate expansion of mortars. (Adapted from Tokyay, M., Dilek, F.T. 2003. Sulfate resistance of cementitious systems with mineral additives. Final Rept of Project No: 1991010, 243 pp. Granted by TÜBİTAK, The Scientific and Technical Research Council of Turkey, Construction and Environmental Technologies Research Grant Committee.)

reached a compressive strength of 20 MPa. Expansion measurements were made until 308 days and the sulphate concentration of the solution was monitored in every 2 weeks and was renewed periodically every 4 weeks.

Eglinton (1988) referred to the studies of D.G. Miller and P.W. Manson of the U.S. Department of Agriculture that started in the early 1930s. Specimens prepared using more than 100 different cements were exposed to sodium and magnesium sulphate solutions in the laboratory and immersed in lake water with 3.3% $MgSO_4$, 1.2% Na_2SO_4 and 0.3% $CaSO_4$ for up to 25 years. Although they investigated many different parameters such as surface coatings, curing methods, air-entraining admixtures and mineral additives, their most important finding was the correlation between the sulphate resistance and C_3A contents of the cements. The most resistant specimens were made from cements with an average C_3A content of 4.4% and with a maximum of 5.4%. The average C_3A content of the least resistant ones was 11.9%. So they proposed that PCs with more than 5.5% C_3A should not be used in sulphate environments. This value is very close to what we use as the limit (5%) for sulphate resisting PCs (ASTM C150 Type V or EN 197-1 Type CEM I-SR5) today.

The required sulphate resistance of a concrete is usually attained by blended cements or incorporating pozzolans or GGBFS in concrete. In fact, ACI recommends the use of natural pozzolan, FA, SF or BFS blended cements (ACI 201, 2001) and EN 197-1 specifies slag cements, CEM III/B and CEM III/C, and pozzolanic cements, CEM IV/A and CEM IV/B as sulphate-resisting cements.

There is a vast amount of research on the influence of mineral admixtures on the sulphate resistance of concrete and almost all of them are stated to follow the test procedure described by ASTM C 1012. Relatively new studies are considered here.

In a study on Mexican natural pozzolans, eight different pozzolans were incorporated in different amounts in the PCs with varying C_3A contents. The sulphate expansions or the mortar bars were compared with those of ASTM Type I, II and V PC mortars at 26 and 52 weeks as illustrated in Figure 11.3. Type I cement failed at 16 weeks, therefore, it is not included in the figure. Type I PC clinkers were used in six of the pozzolanic cements and Type V PC clinkers were used in the other two. C_3A contents of Type I clinkers were 14.3% and 10.4% and those of Type V clinkers were 1.0% and 2.6% (Rodriguez-Camacho and Uribe-Afif, 2002).

It is clear from Figure 11.3 that natural pozzolans are effective in increasing sulphate resistance. In the same study, it was stated that higher pozzolanic activity of the natural pozzolan results in better sulphate resistance.

A similar study was carried out by Şahmaran et al. (2007) in which a natural pozzolan and a low-lime fly ash were used to replace ordinary PC with varying proportions. The ingredient proportions of the cements are given in Table 11.4. The clinker of the ordinary PC and the blended cements was the same and contained 9.8% C_3A and 11.4% C_4AF. A sulphate-resisting

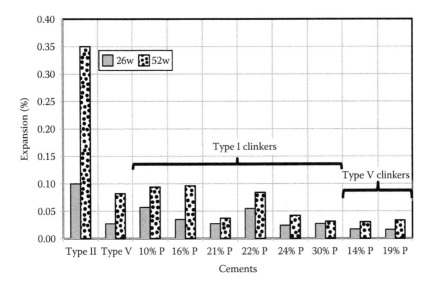

Figure 11.3 Effect of natural pozzolans on sulphate resistance. (Data from Rodriguez-Camacho, R.E., Uribe-Afif, R. 2002. *Cement and Concrete Research*, 32, 1851–1858.)

(ASTM Type V) PC with 3.6% C_3A and 13.8% C_4AF was also used for comparison.

First, seven mortar mixtures were prepared with a constant water-cementitious ratio of 0.485. The blended cement mortars were considerably stiffer than the PC mortars. Then, a second set of mixtures was prepared by adjusting the water–cementitious ratio of the blended cement mixtures to obtain flow values similar to those of the PC mixtures in the first set. The water–cementitious ratios ranged from 0.54 to 0.57 with an average of 0.56. ASTM C 1012 method was applied to all specimens. Ordinary portland cement (OPC) mortars disintegrated at very early ages. Sulphate expansions of the mortar bars at 26, 52 and 78 weeks are shown in Figure 11.4.

Table 11.4 Proportions of the ingredients in ordinary portland cement and blended cements used by Şahmaran et al. (2007)

Cement label	Proportions of the ingredients (%)			
	Clinker	Natural pozzolan	Fly ash	Limestone
OPC	96.5	0	0	3.5
BC	70.8	10.8	14.9	3.5
BC$_{FA}$	64.7	0	31.8	3.5
BC$_{NP}$	66.3	30.2	0	3.5
BC$_{NP-FA}$	61.2	22.2	13.1	3.5
BC$_{FA-NP}$	60.3	15.3	20.9	3.5

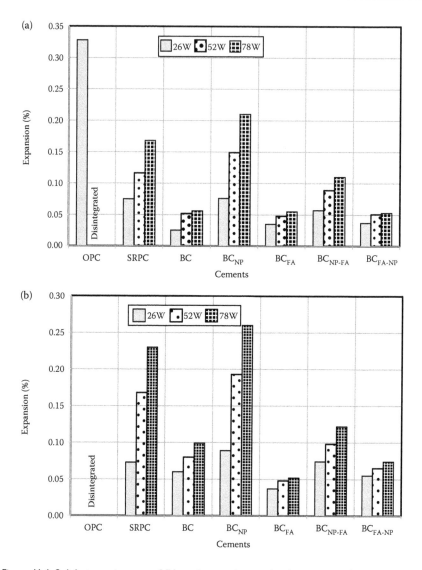

Figure 11.4 Sulphate resistance of FA and natural pozzolan incorporated cement mortars in comparison with ordinary and sulphate resisting PC mortars with water–cementitious ratios of (a) 0.485 and (b) 0.560. (Data from Şahmaran, M., Erdem, T.K., Yaman, I.O. 2007. *Construction and Building Materials*, 21(8), 1771–1778.)

It was concluded that blended cements prepared with FA or natural pozzolan or combinations of both improves the sulphate resistance by (a) reducing the C_3A content and (b) consuming some of the CH produced upon the hydration of PC by pozzolanic reactions. Blended cement prepared by using low-lime fly ash showed the best performance.

The positive effect of incorporating low-lime fly ashes in concrete on the sulphate resistance has long been known (Davis et al., 1937). On the other hand, the effect of high-lime fly ashes still remains controversial. Generally, fly ashes with less than 15% CaO are stated to be improving the sulphate resistance of concrete (ACI 232.2R, 2003). An indicator of the sulphate resistance of a FA was developed by Dunstan (1980). The 'R-value' he proposed is given in Equation 11.11. According to Dunstan, if R < 0.75, sulphate resistance is greatly improved. For R = 0.75–1.50, a moderate improvement is achieved. If R = 1.5–3.00, no significant change occurs and for R > 3.00, sulphate resistance is reduced.

$$R = \frac{\%(CaO)_{fly\ ash} - 5}{\%(Fe_2O_3)_{fly\ ash}} \hspace{3cm} (11.11)$$

Later on, the R-value was criticised by several researchers who stated that presence of reactive alumina and other expansive phases are more significant in affecting the sulphate resistance and Fe_2O_3 does not strongly influence it (Mehta, 1986b; Erdoğan et al., 1992; Tikalsky and Carasquillo, 1993; Thomas et al., 1999). Indeed, Ramyar (1993) found out in his PhD thesis that the reactive alumina and free lime contents of high-lime fly ashes are the two important factors affecting the sulphate resistance. He determined the reactive alumina contents of the fly ashes by rapid HCl treatment. The sulphate expansion data of 18 high-lime fly ash-incorporated cements obtained according to ASTM C 452 (1985) were processed statistically and it was determined that there is a fair relationship between the ratio of (C_3A + reactive alumina)-to-free lime of the FA cements and sulphate expansion, as illustrated in Figure 11.5. Although the sulphate expansion test utilised is not considered to be appropriate for blended cements, it is obvious that the reactive alumina and free lime contents of high-lime fly ashes play an important role on their efficacy under sulphate attack.

Limestone powder, low-lime fly ash, high-lime fly ash, GGBFS, SF and MK were used in a comprehensive experimental study on sulphate attack in mineral admixture-incorporated mortars. Two series of specimens were prepared. Limestone-incorporated cements with 4%, 15% and 22% limestone (by mass) in the first and 4% and 10% limestone in the second series were used. C_3A contents of the cements used in the first and second series were 8%–9% and 11%–12%, respectively. Some of the mixture groups in the first series included varying amounts of low-lime fly ash (15%–35%), GGBFS (20%–60%) and ternary mixtures of FA (20%) or GGBFS (25%) mixed with SF (5%), as partial replacement of limestone-incorporated cements. In the second series, three of the mixture groups contained 12%–18% MK; one contained 8% SF; one contained 25% low-lime fly ash; another one contained 25% high-lime fly ash. Three of the mixture groups used triple blends of 15% + 10% and 20% + 10% of high-lime

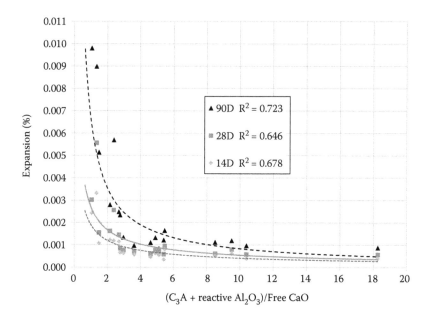

Figure 11.5 Relationship between (C_3A + reactive alumina)-to-free lime ratio and sulphate expansion. (Data from Ramyar, K. 1993. Effects of turkish fly ashes on the portland cement-fly ash systems, PhD thesis, Middle East Technical University, 208pp.)

fly ash and MK and 15% + 5% of low-lime fly ash and SF. Besides this, an ASTM Type V cement with 4% limestone was also used for comparison. All mixtures had the same water–cementitious ratio of 0.485. After attaining 20 MPa strength, half of the specimens were put into 5°C and the other half were put into 23°C 5% Na_2SO_4 solutions and expansions were recorded periodically. Failure was defined either as 0.10% increase in length or disintegration (Hossack and Thomas, 2015). It was concluded that, including mineral admixtures into limestone-incorporated cements improves the sulphate resistance and higher replacement levels may be necessary for higher amounts of limestone powder in the cement. The greatest degree of sulphate resistance was achieved by ternary blends of SF and low-lime fly ash. Deterioration due to sulphate attack is more severe at 5°C than at 23°C, which is attributed to thaumasite formation at lower temperatures. Thaumasite was observed in all Series 1 specimens kept in 5°C sodium sulphate solution.

Irassar (2009), in his extensive review, stated that the sulphate resistance of mixtures made from limestone-incorporated cements is firstly governed by their resistance to sulphate ions penetration, as it is the case for all cementitious systems. Therefore, the water–cementitious ratio is the major factor in controlling sulphate attack. Lower proportions (<10%) of limestone powder do not cause significant reduction in sulphate resistance but higher proportions (>15%) may result in worse sulphate attack damage than the control PC mixtures. Deterioration caused by thaumasite occurs in mixtures that were previously damaged by ettringite formation.

The sulphate resistance of GGBFS-incorporated concrete depends on C_3A content of the PC fraction and the alumina content of the slag (Moranville-Regourd, 1988). It is generally accepted that sulphate resistance is increased with increase in slag content. However, blended cements with GGBFS levels of at least 50% are necessary for a sulphate resistance comparable to that of a sulphate resisting PC. On the other hand, when used at levels of 20%–50%, slag with high alumina content (17.7%) reduced the sulphate resistance whereas slag with lower alumina content (11%) increased the sulphate resistance when compared with ordinary PC (Locher, 1966). Osborne (1999) recommended that for providing good sulphate resisting properties, GGBFS should have alumina contents less than 14%. Otherwise, the C_3A content of the PC fraction should not be higher than 10%.

Where BFS is used in sufficient amounts, improved sulphate resistance is attributed to (a) C_3A dilution, (b) reduction of calcium hydroxide and (c) reduced permeability (ACI 233, 2003).

The positive effect of SF incorporation on sulphate resistance was already stated above for the study of Hossack and Thomas (2015). Khan and Siddique (2011) cited ten investigations on the effect of SF on sulphate resistance, in their review article. In all of those studies SF was found beneficial. However, unlike other pozzolanic materials, the effect is not due to C_3A dilution of the PC fraction since the amount of SF used is much smaller, but more due to the reduced permeability that makes the ingress of external sulphate ions into concrete difficult (ACI 234, 2000).

Several researchers have studied the effectiveness of mineral admixtures on DEF and concluded that the risk of DEF is reduced by mineral admixture incorporation in concrete (Odler and Chen, 1996; Miller and Conway, 2003; Ramlochan et al., 2003, 2004; Tosun, 2007; Adamopoulou et al., 2011; Nguyen et al., 2013). The positive effect of FA, natural pozzolans and GGBFS is attributed to the alumina present in these materials which is stated to readily react with the sulphate ions in the pore solution during the heating process to result in monosulphoaluminate hydrate precipitation. Thus, the remaining sulphate ion concentration becomes insufficient for DEF (Taylor et al., 2001; Tosun, 2007). This may be the explanation of the relatively lower effectiveness of SF, which does not contain alumina (Tosun, 2007).

11.3 SEA WATER ATTACK

Marine environments may be aggressive for concrete. There are several aspects of concrete deterioration due to sea water attack and they generally act in a combined manner. The presence of about 3.5% soluble salts in seawater forms the basis of its chemical attack on concrete. Different types

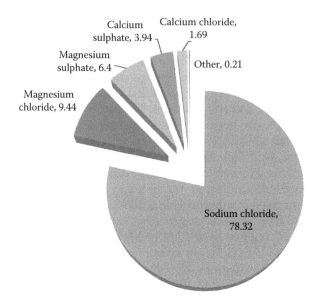

Figure 11.6 Typical relative proportions of salts in seawater, %. (Data from Biczok, I. 1967. *Concrete Corrosion and Concrete Protection*. Chemical Publishing Company, New York, NY; [Ed. T.Y. Erdoğan], 2007. *Beton* [2nd Ed.] METU Press, Ankara.)

of salts in their relative proportions in seawater are illustrated in Figure 11.6. The Na^+, Cl^-, Mg^{2+} and SO_4^{2-} concentrations are sufficiently high in seawater for both sulphate attack and reinforcement corrosion.

A list of major ionic species and their concentrations in various sea waters are given in Table 11.5. It should be noted that the effect of seawater may not be only limited to sulphate attack or reinforcement corrosion. Alkali–aggregate expansion when concrete contains reactive aggregates

Table 11.5 Typical major ion concentrations in various sea waters

	Concentration (ppm) in						
Ion	Eastern Mediterranean Sea	Black Sea	Central Pacific Ocean	South Atlantic Ocean	Persian Gulf	Caribbean Sea	World (average)
Na^+	11,560	4,900	10,700	10,779	12,950	11,035	11,000
K^+	420	230	381	386	nd	397	400
Mg^{2+}	1,780	640	1,290	1,293	1,580	1,330	1,330
Ca^{2+}	470	240	432	421	440	418	430
SO_4^{2-}	3,060	1,360	2,280	2,709	3,200	2,769	2,760
Cl^-	21,380	9,500	26,000	19,370	23,922	19,841	19,800
HCO_3^-	162	nd	175	141	122	146	–

Source: Adapted from Erdoğan, T.Y. 2007. *Beton* (2nd Ed.), METU Press, Ankara; Aydın, F., Ardalı, Y. 2012. *Sigma 30 Journal of Engineering and Natural Sciences*, 156–178.

and carbonic acid effect due to the presence of dissolved CO_2 in shallow waters should also sometimes be considered. Besides chemical attack, physical deterioration caused by wetting and drying, freezing and thawing and abrasion due to wave action may aggravate undesirable effects.

One of the main deteriorative effects on concrete in marine environments is sulphate attack. However, deterioration of concrete in seawater does not show the same characteristics of the deterioration due to sulphate attack by non-saline solutions. It is commonly observed that the sulphate attack of saline water results in erosion and mass loss type of deterioration rather than the disruptive expansion commonly observed in classical sulphate attack. It seems that ettringite expansion is retarded by the presence of chloride (Eglinton, 1988). This behaviour is consistent with the conjecture which states that an alkaline environment is necessary for the expansion of ettringite by water adsorption (Mehta and Monteiro, 2006).

The reciprocal influence of chlorine and sulphate ions on concrete was studied recently by Maes and De Belie (2014). They compared four concretes with same mix proportions but different cements. The cements used were CEM I 52.5N (OPC), CEM I 52.5 N HSR (HSRPC) and two BFS cements with 50% and 70% GGBFS blended with ordinary PC (SC50 and SC70, respectively). To determine the chloride penetration resistance and the effect of sulphates, a diffusion test was carried out and chloride diffusion coefficients were determined. For determining the extent of sulphate attack and the influence of chlorides, the length and mass changes of the specimens were measured. The test solution (Cl) used for chloride penetration resistance contained 165 g/L NaCl and two more solutions (Cl + S1) and (Cl + S2) with 27.5 g/L and 55.0 g/L Na_2SO_4, respectively, added to the first were used to determine the effect of sulphates. For sulphate attack, they used a 50.0 g/L Na_2SO_4 (S) solution and to determine the effect of chloride 50 g/L NaCl was added to this solution (S + Cl).

The diffusion coefficients of slag cement concretes were found to be significantly smaller than those of PC concretes and chlorine ion diffusion reduces with age, as illustrated in Figure 11.7. Consequently, chloride penetration depths at later ages were found to be less in slag cement concretes. The influence of the presence of sulphates in the chloride solution seems to be insignificant from the chloride penetration point of view.

In the same study (Maes and De Belie, 2014), ordinary PC specimens showed the least resistance to sodium sulphate attack. Slag cement specimens were more sulphate resistant than the high-sulphate resistance cement specimens. The presence of sodium chloride in the sulphate solution was found to improve the sulphate resistance.

Another important finding of Maes and De Belie's investigation was the formation of Friedel's salt ($Ca_2Al(OH)_6Cl\cdot2H_2O$) in the specimens immersed in NaCl solution which indicates chloride binding. In addition, similar behaviour was observed by Barberon et al. (2005) through multinuclear magnetic resonance. Maes and De Belie (2014) determined that slag

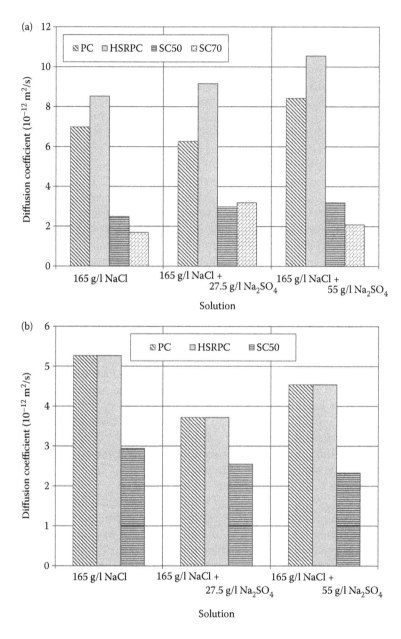

Figure 11.7 Non-steady-state diffusion coefficients of specimens immersed in chloride and chloride + sulphate solutions (a) at 28 days and (b) 84 days. (Data from Maes, M., De Belie, N. 2014. *Cement and Concrete Composites*, 53, 59–72.)

cement specimens had relatively more Friedel's salt than the PC specimens, as measured by quantitative XRD and concluded that the presence of chloride ions has a mitigating effect on the sodium sulphate attack on concrete. Zuquan et al. (2007) stated that the aluminous component of the FA has the ability to react with chlorine ions to produce Friedel's salt, therefore, FA incorporation in concrete has a chloride binding effect. Since it is the free chlorides that initiate the reinforcement corrosion this chloride binding ability stated by many researchers is significant.

Zhang et al. (2013) found out that presence of chloride ions in concrete pore solution reduces sulphate attack. Similar behaviour was observed by Sotiriadis et al. (2012, 2013) who studied the effect of chloride on thaumasite formation in concrete mixtures prepared with portland-limestone cements with and without several different mineral admixtures partially replacing it. They used a high-lime fly ash (30%), a GGBFS (50%), a natural pozzolan (30%) and an MK (10%) to replace portland-limestone cement which contained 15% (by mass) limestone powder. Deterioration was found to be substantially lower in chloride–sulphate solutions than in sulphate solutions alone and mineral admixture incorporation further improved the resistance.

Use of pozzolanic materials had been also shown to improve the resistance of concrete against seawater attack by many other researchers (Bai et al., 2003; Shannag and Shaia, 2003; Hossain, 2008; Donatello et al., 2013).

Generally, sulphate exposure by seawater is considered as moderate (Class 01 in Table 11.2) and the use of medium sulphate-resisting types of cements (MS), for example ASTM C 150 Type II, is found to be appropriate (ACI 318, 2008). The reason why seawater attack is considered as less intense than one would assume from its sulphate and magnesium ion contents is that HCO_3^- and Mg^{2+} ions in the seawater form highly insoluble calcium carbonate and magnesium hydroxide upon reacting with the hydration products and these compounds precipitate in the pores and thus slow down the further ingress of sulphates as well as chlorides (Massazza, 1988).

In designing concrete mixtures against sulphate attack by seawater, the water–cementitious ratio is of prime significance as for many other durability issues. It is required that the water–cementitious ratio should be at most 0.50. In selecting a cement, C_3A content is the principal consideration as for all cases for sulphate resistance. It is generally accepted that the C_3A content of the PC should be less than 8% for moderate sulphate exposure. However, if the water–cement ratio is lowered to 0.40, use of PCs with up to 10% C_3A is permitted (ACI 318, 2008).

11.4 CARBONATATION

Hardened cement paste may react chemically with atmospheric CO_2. The CO_2 diffuses into concrete, dissolves in the pore solution and then reacts with various calcium-bearing hydration products to form carbonates. The term *carbonatation* is preferred in this book to the commonly used term *carbonation* since the reactions result in *carbonated* products. The main carbonatation reaction is with calcium hydroxide, which is one of the major hydration products of PCs:

$$Ca(OH)_2 + CO_2 \rightarrow CaCO_3 + H_2O \uparrow \qquad (11.12)$$

However, carbonatation is not only limited to calcium hydroxide. C–S–H gel, unhydrated C_3S and C_2S particles, KOH, NaOH, $Mg(OH)_2$ and calcium aluminate phases may also carbonate (Peter et al., 2008). The carbonatation of non-calcium bearing compounds and calcium aluminate phases are considered to be not as significant as those of calcium hydroxide and calcium silicate compounds from the durability and dimensional stability points of view. Carbonatation reactions of C–S–H gel and any remaining C_3S and C_2S can be described, in terms of cement chemistry abbreviation, by the following chemical equations, respectively:

$$C - S - H + \bar{C} \rightarrow C - S - H + C\bar{C} + H \qquad (11.13)$$

$$C_3S + 3\bar{C} + H \rightarrow SH + 3C\bar{C} \qquad (11.14)$$

$$C_2S + 2\bar{C} + H \rightarrow SH + 2C\bar{C} \qquad (11.15)$$

Note that Equations 11.14 and 11.15 describe also the carbonatation of PCs during prolonged storage by picking up CO_2 and humidity from the atmosphere which results in some loss of cementitious value of the cement.

Carbonatation may have two significant deteriorative consequences: (1) shrinkage and (2) risk of reinforcement corrosion due to reduced alkalinity of the concrete. The former is commonly thought to be caused by the loss of water obtained as a result of carbonatation of calcium hydroxide, by evaporation (Equation 11.12) and generally accepted to be not so serious since the expansion upon calcium carbonate formation and shrinkage by the loss of water would more or less compensate each other. However, when Equation 11.13 is considered, it can be seen that the lime–silica ratio of the C–S–H gel reduces upon carbonatation and this changes both the structure of the gel and its binding characteristics (Mindess and Young, 1981). The change in the structure of the C–S–H is caused by the decomposition of the silicate anion structure that leads to shrinkage (Matsushita et al., 2004). So, carbonatation may result in shrinkage of concrete that would normally be reached at much lower relative humidities. Furthermore, the shrinkage caused by the carbonatation of the C–S–H gel is completely irreversible (Mindess and Young, 1981). On the other hand, it was observed in autoclaved aerated concrete through 29Si MAS NMR spectroscopy that a critical level of carbonatation (~20%) should be reached by the C–S–H gel in order for it to shrink (Matsushita et al., 2004).

Carbonatation may reduce the pH of concrete from values around 13 to values as low as 8. The importance of the lowered alkalinity of concrete is related to the fact that it increases the risk of reinforcement corrosion. Carbonatation-induced corrosion is affected by several factors such as concrete cover thickness, degree of compaction, degree of hydration and porosity. Ingress of CO_2 into concrete is related with these factors. Carbonatation proceeds into concrete roughly following the diffusion law, which can be defined as the rate being inversely proportional to the thickness:

$$\frac{dx}{dt} = \frac{D_0}{x} \tag{11.16}$$

where x is the distance, t the time and D_0 the diffusion constant.

D_0 is dependent on the concrete quality. The diffusion is easier in concretes with more open pore structure. Therefore, good compaction and curing are necessary.

There are a number of empirical equations proposed to estimate the carbonatation depth (Broomfield, 1997). The basic equation can be given as

$$d = At^n \tag{11.17}$$

where d is carbonatation depth, t the time, A the diffusion coefficient which generally ranges from 0.25 to 1.00 mm/year$^{0.5}$ depending on concrete quality, and n is constant (0.50).

The influence of mineral admixtures on the carbonatation of concrete is not yet fully understood. Results obtained from different experimental investigations tend to vary considerably, probably because of the differences in concrete mix proportioning and the accelerated test methods employed. It should be noted that in almost all of the experimental investigations on concrete carbonatation, CO_2 concentrations much higher than those can be encountered under actual atmospheric conditions are used. Average global CO_2 concentration in atmosphere is 0.04% whereas concentrations upto 50% were used in the tests to accelerate the carbonatation.

Generally, the positive influence of mineral admixtures is attributed to the reduced porosity whereas the negative influence is attributed to the lowered alkalinity due to pozzolanic reactions.

In a recent study by Younsi et al. (2013) concretes of approximately the same 28-day compressive strength were prepared by using three commercially available European cements, CEM I (PC), CEM II/BV (portland fly ash cement) and CEM III/C (slag cement) and four laboratory cements prepared by blending 30% and 50% (by mass) FA and 30% and 75% (by mass) GGBFS with CEM I. CEM II/BV and CEM III contained 23% FA and 82%

BFS, respectively. Two sets of specimens were prepared and one set was continuously moist cured and the other set was air-cured until the time of carbonation test. Porosities of the specimens were measured to be similar for all mixes under the same curing. Obviously, air-cured concretes had more porosity than their moist-cured counterparts. The specimens were put in the carbonatation chamber containing $50\% \pm 5\%$ CO_2 at 20°C and 65% RH. Carbonatation depth measurements started after 28 days and continued for 123 days. CEM I had the least and CEM III/C had the most carbonatation and the depth of carbonatation increased with increasing mineral admixture content. Moist curing improves the carbonatation resistance and it is more pronounced in FA- and slag-incorporated concretes.

Similarly, the effect of high-volume FA incorporation on carbonatation was studied by Atiş (2003) using moist- and air-cured (65% RH) specimens. FA amounts used were 50% and 70% of the total cementitious mass. A carbonatation test was carried out at about 5% CO_2 concentration for 2 weeks. Although 70% FA incorporation resulted in higher carbonatation, 50% FA incorporation resulted in considerably less carbonatation than the control for moist-cured specimens and almost the same carbonatation as the control for air-cured specimens, as illustrated in Figure 11.8.

Bai et al. (2002) determined that as the amount of FA replacing the PC increases carbonatation depth also increases. On the other hand, partially replacing the FA with MK was found to be beneficial.

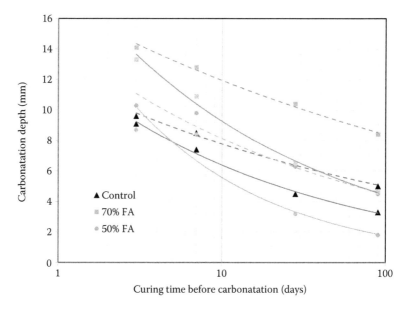

Figure 11.8 Carbonatation depths after 2 weeks in portland cement and FA-incorporated concretes. Full lines indicate moist-cured and dashed lines indicate air-cured concretes. (Data from Atiş, C.D. 2003. *Construction and Building Materials*, 17, 147–152.)

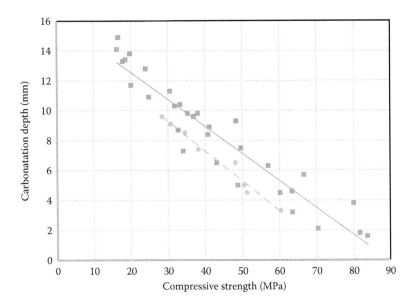

Figure 11.9 Strength-carbonatation depth relationship of portland cement (circle points) and FA incorporated concretes (square data points). (Data from Atiş, C.D. 2003. *Construction and Building Materials,* 17, 147–152.)

High-lime fly ash was determined to be more effective in increasing the carbonatation resistance of concrete than low-lime fly ash when used to either partially replace the PC or sand (Papadakis, 2000). SF incorporation generally results in higher carbonatation depths than the control concretes without it (Papadakis, 2000; Khan and Siddique, 2011; Torgal et al., 2012).

Lollini et al. (2014) replaced 15% and 30% (by mass) PC with limestone powder and determined that the carbonatation resistance does not change significantly with respect to the control in the former but is reduced in the latter.

The results of the experimental investigations on the carbonatation of mineral admixture incorporated concretes vary considerably. However, it is generally accepted by many researchers that if two concretes with and without mineral admixtures are designed to have the same workability and strength, then it is most probable that they will carbonate to a similar depth. Indeed, the strength and carbonatation depth data of Atiş (2003) are plotted to show that FA incorporated concretes and PC concretes having similar strengths also have similar carbonatation depths (Figure 11.9).

11.5 REINFORCEMENT CORROSION

When two metals of different electrochemical potentials are coupled in the presence of moisture and oxygen, electrochemical corrosion can occur.

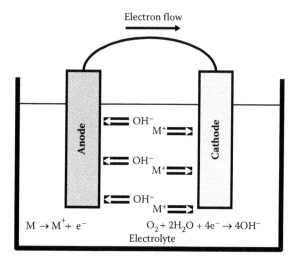

Figure 11.10 Simple schematic representation of electrochemical corrosion.

A typical corrosion cell is illustrated in Figure 11.10. Considering the anode as iron, the following reactions occur at the anode and the cathode:

$$Fe - 2e \rightarrow Fe^{2+} \quad \text{(at the anode)} \tag{11.18}$$

$$2H_2O + O_2 + 4e^- \rightarrow 4(OH)^- \quad \text{(at the cathode)} \tag{11.19}$$

Then, rust forms as the ferrous oxide turns into hydrated ferric oxide:

$$2Fe + 2H_2O + O_2 \rightarrow 2Fe(OH)_2 \tag{11.20}$$

$$2Fe(OH)_2 \xrightarrow{O_2,H_2O} 2Fe(OH)_3 \rightarrow Fe_2O_3 \cdot nH_2O \tag{11.21}$$

Corrosion of steel embedded in concrete is also an electrochemical process. Although there is no separate cathodic metal present, different areas of steel may have different electrochemical potentials resulting in anodic and cathodic portions forming couples. Moisture within the concrete acts as an electrolyte. Thus a corrosion cell is formed as illustrated in Figure 11.11.

The possible anodic and cathodic reactions that may occur in the reinforcement steel are as follows (Ahmad, 2003):

$$3Fe + 4H_2O \rightarrow Fe_3O_4 + 8H^+ + 8e^- \quad \text{(at the anode)} \tag{11.22}$$

$$2Fe + 3H_2O \rightarrow Fe_2O_3 + 6H^+ + 6e^- \quad \text{(at the anode)} \tag{11.23}$$

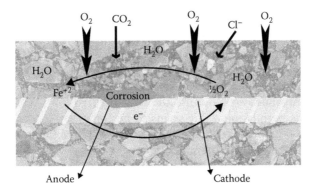

Figure 11.11 Schematic representation of reinforcement corrosion in concrete.

$$Fe + 2H_2O \rightarrow HFeO_2^- + 3H^+ + 2e^- \quad \text{(at the anode)} \tag{11.24}$$

$$Fe \rightarrow Fe^{2+} + 2e^- \quad \text{(at the anode)} \tag{11.25}$$

$$2H_2O + O_2 + 4e^- \rightarrow 4(OH)^- \quad \text{(at the cathode)} \tag{11.26}$$

$$2H^+ + 2e^- \rightarrow H_2 \quad \text{(at the cathode)} \tag{11.27}$$

The electromotive force (emf, ε) of the reinforcement corrosion cell formed can be written as

$$\varepsilon = 0.0148 \log(O_2) - 0.0591(\text{pH}) - 0.0296 \log(Fe^{2+}) + 1.669 \, [\text{V}] \tag{11.28}$$

where, O_2 and Fe^{2+} are expressed as concentrations in the electrolyte (capillary water).

It is evident from Equation 11.28 that the reinforcement corrosion rate is affected by the availability of oxygen and capillary water and the pH of the concrete (Ahmad, 2003). There are numerous internal and external factors such as the properties of concrete making materials, concrete quality, cover over reinforcement, water–cement ratio, permeability, mix proportions, workmanship, ambient temperature and humidity and curing that may affect these two parameters and thus the corrosion of reinforcement in concrete.

Concrete, with its high alkalinity, usually provides sufficient protection against reinforcement corrosion. A thin but dense layer of iron oxide which is called the *passive layer* forms on the surface of the reinforcement. This layer prevents further corrosion. However, when the pH is lowered from 13 to 10 or below corrosion can occur. Carbonatation, as discussed in the previous section, can lower the pH of concrete considerably. A strong electrolytic solution is necessary to accelerate corrosion. The presence of chloride ions serve that purpose and promote reinforcement corrosion. A maximum of 0.20% (by mass of cement) acid soluble chloride ion content in a concrete mix is recommended to minimise the risk of corrosion (ACI 201, 2001). Although this permissible value seems to be conservative, considering the possibility of regeneration of chloride ions as given by Equations 11.29 and 11.30, it is better to be on the safe side.

$$Fe^{3+} + 3Cl^- \xrightarrow{\text{electrolyte}} FeCl_3 \tag{11.29}$$

$$FeCl_3 + 3(OH)^- \rightarrow Fe(OH)_3 + 3Cl^- \tag{11.30}$$

Besides CO_2 and Cl^-, of course, moisture and supply of oxygen must be present. Concrete reinforcement corrosion is generally represented by two stages: (a) initiation and (b) propagation, as illustrated in Figure 11.12. The first stage requires carbonatation and presence of free chloride ions and the second stage requires humidity and oxygen diffusion.

The effects of mineral admixtures on chloride ion diffusion and carbonatation were already discussed in the previous two sections. Various

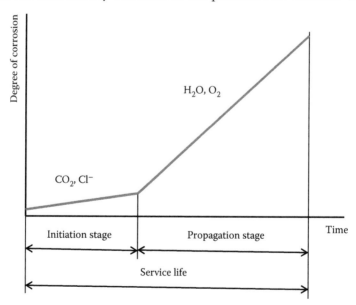

Figure 11.12 Schematic presentation of reinforcement corrosion in concrete. (Adapted from Tuutti, K. 1982. Corrosion of Steel in Concrete. CBI Research 4.82, Swedish Cement and Concrete Research Institute, Stockholm.)

investigations related with their effect on reinforcement corrosion will be considered here.

FA had been shown to be effective in reducing the risk of reinforcement corrosion by many researchers (Cabrera, 1996; Thomas, 1996; Montemor et al., 2000, 2002; Angst et al., 2011).

Thomas (1996) determined the mass loss of reinforcements and the chloride contents of the concretes at the location of reinforcement in reinforced concrete specimens exposed to tidal conditions for 1–4 years for three different strength grades and four different (0%–50%) FA contents. He found out that the threshold chloride level that could be tolerated without significant mass loss in the reinforcement decreased with increased FA content. Still, however, FA concretes were stated to provide better protection of steel due to their increased resistance to chloride ion penetration.

Montemor et al. (2000) studied the reinforcement corrosion in concretes with 0%, 15%, 30% and 50% (by mass) FA replacing the PC in 3% NaCl solutions under full and partial immersion conditions by electrochemical impedance spectroscopy (EIS). They have found out that FA is beneficial both at the initiation and propagation stages of corrosion. The beneficial effect is attributed to (1) the chloride-binding ability of the FA by forming chloroaluminates and thus reducing the free chloride content; (2) thickening of the passive film in the presence of FA and finally (3) decreased porosity by both physical and pozzolanic means. In another study, Montemor et al. (2002) carried out a similar investigation this time under the combined action of CO_2 and NaCl. Under natural carbonatation conditions, they have obtained results similar to the previous study. However, under accelerated carbonatation conditions where CO_2 concentration is much higher than the natural level, chloride contents were found to be higher in FA incorporated specimens than the control specimens. The reason for this adverse effect was stated as the destruction of the chloroaluminates by the attack of CO_2 which reveals more free chloride ions. Therefore it is suggested that care should be taken when accelerated tests are used to estimate the behaviour under natural conditions.

PC replacement by FA was determined beneficial regarding the chloride penetration, resistance and electrical resistivity of concrete. However, FA incorporated concretes were found more vulnerable to chloride exposure at early ages (Angst et al., 2011). It should be noted, however, that the FA concretes had lower 28-day strengths than the control concretes, in the study by Angst et al.

A review by Song and Saraswathy (2006) who discussed the studies on the reinforcement corrosion resistance of GGBFS incorporated concretes concludes that (a) the reduction of pH value of concrete due to GGBFS incorporation does not have an adverse effect; (b) corrosion rate of reinforcement does not change significantly by replacing the PC up to 40% GGBFS but at higher levels of replacement it is reduced considerably and (c) adequate curing is essential to reduce the susceptibility of GGBFS concrete against carbonatation.

In a more recent study reinforcement corrosion was investigated under an organic acid environment which resembles the conditions that may be encountered in agricultural structures. Specimens prepared using ordinary PC, 80% slag cement and 20% MK cement had the same water–cementitious ratio of 0.65. All specimens were moist cured for 6 months and then tested to measure the corrosion potential and polarisation resistance for determining the probability of and initiation of corrosion in acetic acid under repeated wetting and drying conditions (4 days wetting and 3 days drying) for 429 days. Corrosion initiation was earlier in GGBFS specimens than PC specimens. MK specimens had almost a twice as long corrosion initiation period than the control specimens (Oueslati and Duchesne, 2012).

Studies on the effect of SF on reinforcement corrosion revealed that concretes containing SF have better corrosion resistance than their control counterparts. When used to partially replace both the PC and fine aggregate, 6% and 12% SF resulted in 2.5 and 5 times more electrical resistivity of concrete than that of the control, respectively. The time to start the reinforcement corrosion was also determined to be longer in SF concretes (Dotto et al., 2004). High-strength concrete containing 10% (by mass of PC) SF showed extremely low values of corrosion current density even after very long exposure to chloride solution (Kayali and Zhu, 2005). Similarly, concretes containing SF to replace 20% (by mass) PC had significantly lower corrosion than the corresponding concretes without SF (Cabrera et al., 1995). The studies mentioned here and many more attribute the better corrosion resistance of SF concrete to its lower permeability.

11.6 ALKALI–AGGREGATE REACTIVITY

Alkali–aggregate reactivity (AAR) is broadly defined as the chemical reactions between various components of aggregates and the alkalies in the cement. When reactive siliceous aggregates are involved, it is called alkali–silica reactivity (ASR) and when reactive carbonate aggregates are involved it is called alkali–carbonate reactivity (ACR). The former is more common than the latter. Although the mechanisms of reactions in ASR and ACR are totally different from each other, both reactions may lead to disruptive expansion.

Concrete pore water contains sodium (Na^+), potassium (K^+), calcium (Ca^{2+}) and hydroxyl (OH^-) ions. Concentrations of the first three ions depend on their original amounts in the unhydrated cement. Generally, pH of the pore water is 12.7–13.1 in low-alkali cements and may be as high as 13.9 in high-alkali cements. Increase in the pH of the pore solution is an indication of the solubility of amorphous silica and its rate of dissolution. Approximate solubility of amorphous silica is around 100 ppm in neutral water which has a pH of around 7. It increases to about 500,000 ppm in cement paste with a pH of 12.5 and becomes almost completely soluble in a high-alkali cement paste with pH > 13 (Mindess and Young, 1981).

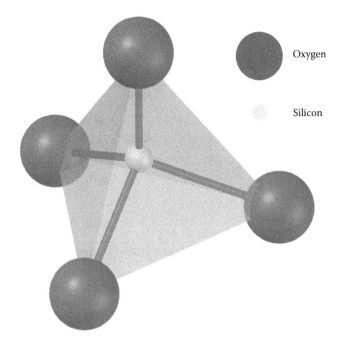

Figure 11.13 Silica tetrahedron.

The silica structure is represented by a tetrahedron which constitutes a Si^{4+} at the centre and four oxygen ions (O^-) at the corners, as illustrated in Figure 11.13. In crystalline silica such as quartz, the tetrahedra are linked to each other through the bonding of each oxygen atom to two silicons, forming a regular structure which is chemically and mechanically stable. Only the surface of crystalline silica may be reactive since complete tetrahedra cannot form there. In amorphous silica, however, the tetrahedra are arranged in a random form, resulting in a more porous structure. Crystalline and amorphous structures of silica are shown schematically in Figure 11.14.

The chemistry involved and the mechanism of ASR are rather sophisticated and several different approaches were made so far for its explanation. Nevertheless, it can be described in three steps, in a simplified manner as follows (Glasser, 1992; Andiç-Çakır, 2007; Ichikawa and Miura, 2007):

The first step of ASR is the cutting of siloxane network by the action of hydroxyl ions:

$$-Si - O - Si -+R^+ + OH^- \rightarrow -Si - O - R + H - O - Si - \quad (11.31)$$

where R^+ is Na^+ or K^+.

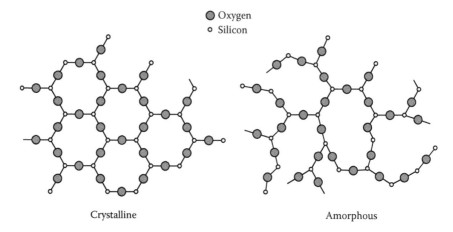

Figure 11.14 Two-dimensional representation of crystalline and amorphous silica structures.

The silicic acid obtained in Equation 11.31 reacts further to convert into alkali silicate in the second step:

$$H-O-Si-+R^{+}+OH^{-} \rightarrow -Si-O-R+H_2O \qquad (11.32)$$

Finally, in the third step, the resulting alkali silicate which is hygroscopic, imbibes water and expands:

$$-Si-O-R+nH_2O \rightarrow -Si-O-R\cdots(H_2O) \qquad (11.33)$$

Alkali silica gel has an unlimited moisture absorption capacity. The volume expansion generates internal stresses that cause cracking of the aggregate and the matrix surrounding it as they exceed the tensile strength. The local cracks then interconnect with each other and eventually the concrete fails.

Expansion caused by ASR was determined to be a function of the ratio of the reactive silica-to-alkali concentration. This brings about the concept of *pessimum percentage* which is defined as the reactive silica content of the aggregate that results in the maximum expansion upon ASR. For certain critical values of reactive silica in the aggregate, the alkali content is sufficient to interact with those particles and result in local disruptive expansions. However, when the reactive aggregate content gets more than the critical amount, the alkali concentration of the concrete will not be sufficient to result in the complete reaction of all particles and thus, expansion reduces (Mindess and Young, 1981; Hobbs, 1988; Merriaux et al., 2003). Maximum expansion occurs when the reactive silica–alkali ratio is around 4.5 (Hobbs, 1988).

The size of the reactive aggregate particles was found to be another factor affecting ASR (Diamond and Thaulow, 1974; Hobbs and Gutteridge, 1979; Zhang et al., 1999; Ramyar et al., 2005). Although there are contradictory results reported in the literature on the aggregate size effect due to the differences in the types and amounts of the reactive aggregates used in different studies, it can generally be concluded that small particle sizes cause rapid reaction without deleterious effects.

Obviously, the alkali content of concrete is another prime factor affecting the ASR. The majority of alkalies in PC concretes come from the cement although various aggregates such as feldspars, zeolites, clay and sea sand may contribute to the total alkali amount. The alkali content of the portland cement clinker is directly related with those of the raw materials used. Therefore, it may sometimes be difficult to manufacture low-alkali cements. Furthermore, stringent environmental regulations may also hinder the production of low-alkali cements since they cause more use of natural resources and energy and increase cement kiln dust generation.

It is customary to give the amount of alkalis in cement and concrete in terms of an equivalent sodium oxide value which is calculated by Equation 11.34:

$$(Na_2O)_{eq} = Na_2O + 0.658K_2O \tag{11.34}$$

This value is required as maximum 0.6% in the cements that will be used with reactive aggregates.

Expansion and cracking caused by ASR is observed in concretes that contain more than 3.0 kg/m^3 alkalis. Concretes of the same alkali concentration may show varying expansions due to different (a) alkali release rates and (b) sodium–potassium ratios of the cements used and (c) different rates of strength development of the concretes (Hobbs, 1988).

Besides the factors stated above, ambient humidity and temperature also affect ASR. Expansion caused by ASR increases with the increase in relative humidity (Stark, 1991; Poole, 1992) and the rate of reaction increases with increase in temperature (Andiç-Çakır, 2007).

To prevent ASR expansion, one or more of the following three strategies are recommended (ACI 221, 1998).

1. Control of the moisture available
2. Control of the type and amount of reactive siliceous aggregate
3. Lowering the pH of the concrete pore fluid

The expansion mechanism of ACR is even more uncertain. The reaction involves various dolomitic limestones which contain dolomite crystals distributed in the fine-grained calcite and clay matrices (Andiç-Çakır, 2007). ACR may be described by the following simplified chemical reactions:

$$CaMg(CO_3)_2 + 2ROH \rightarrow Mg(OH)_2 + CaCO_3 + R_2CO_3 \qquad (11.35)$$

$$R_2CO_3 + Ca(OH)_2 \rightarrow 2ROH + CaCO_3 \qquad (11.36)$$

where R represents Na and K.

The expansions caused by ACR is attributed to one or more of the following phenomena described by different researchers: clay and other colloidal matter present in the structure of the aggregate is exposed upon the dedolomitisation reaction and absorb water to expand (Feldman and Sereda, 1961; Gillot, 1986), or osmotic pressure develops in the clay membrane surrounding the dolomite crystals (Tong and Tang, 1999), or crystallisation of brucite ($Mg(OH)_2$) and calcite generates pressure (Tang et al., 1986), or cryptocrystalline quartz in the matrix or carbonates reacting with alkalis as in ASR (López-Buendia et al., 2006; Grattan-Bellew et al., 2010; Katayama, 2010).

ACR expansion is affected by similar factors as ASR: increase in (1) the alkali concentration of concrete, (2) the pH of the pore solution, (3) reactive phase content of the aggregate, (4) relative humidity, (5) ambient temperature, (6) size of the aggregate and (7) lower concrete strength result in higher expansion caused by ACR (Andiç-Çakır, 2007).

The control of moisture, control of reactive aggregate and lowering the pH of the pore solution are also good for preventing ACR damage. Besides lowering the size of the reactive aggregate and using cement with alkali contents lower (0.4%) than that is recommended for ASR are stated to be beneficial (ACI 221, 1998).

There are numerous test methods to evaluate the potential or the extent of AAR in concrete and continuous efforts are being made towards improving the methods or developing new ones. The reader may be referred to elsewhere (Wigum, 1995; Grattan-Bellew, 1997; ACI 221, 1998; Duyou et al., 2006; Thomas et al., 2006; Andiç-Çakır, 2007; Lingård et al., 2010) for a comprehensive list and the review of the current test methods.

The use of pozzolans as a measure to prevent ASR had been studied extensively since the 1950s. Since its first reported occurrence by Stanton (1940) many researches were carried out on the effects of reducing or preventing ASR by the appropriate use of natural pozzolans, calcined pozzolans, fly ashes, GGBFS and SF in concrete. The mechanism by which these materials lower the potential of ASR changes with their type, quality and amount. Although the mechanism is not yet fully understood, reduction in ASR upon pozzolan-incorporation is generally attributed to (1) lower permeability and therefore lower ion mobility, (2) lowered pH of the pore solution, (3) lower calcium hydroxide content and (4) higher alkali entrapment by the pozzolanic reaction products. Calcium hydroxide content is reduced by pozzolanic reactions thus causing a lower pH of the pore solution. Moreover, the lower lime–silica ratio of the C–S–H produced by pozzolanic reactions has higher ability to combine alkalis. Both of these phenomena result in lowered pH of the pore solution (Massazza, 1981). Some sort of dilution effect may also be accounted for in describing the effect of mineral admixtures when they are used to partially replace PC: Generally, these materials being less reactive than PC will release less alkalis over a longer period of time. Thus, for a given water–cementitious ratio, the total alkali concentration of the pore water is less in concretes made with blended cements (Glasser, 1992).

Monteiro et al. (1997) have prepared mortar bars from two PCs containing different amounts of three natural pozzolans, an FA and a GGBFS. The PCs and the mineral admixtures used had total alkali contents as 1.37%, 0.68%, 0.73%, 0.10%, 4.58%, 1.85% and 0.54%, respectively. The tests were carried out according to ASTM C 1260. When used to partially replace the higher alkali PC, for a replacement level of 20%, all three natural pozzolans behaved in a very similar manner and reduced the expansion from around 0.15% to less than 0.1% at 30 days. Although the natural pozzolans had different amounts of alkali (0.10%, 0.73% and 4.58%) their effects on the expansion were not much different. This was attributed to the limited release of alkalis from the pozzolans. As the amount of natural pozzolan increases the expansion is significantly reduced. 15% replacement of higher alkali PC by FA did not result in any improvement in reducing ASR expansion. However, significant reductions were experienced for 25% and 30% FA contents (less than 0.08% and 0.05%, respectively). Similarly, only a limited improvement was observed when 50% GGBFS was used but greater reductions in expansion were obtained for 60% and 70% GGBFS contents. Different amounts of GGBFS were also used with lower alkali PC. GGBFS replacement levels up to 45% did not reduce the expansions but for GGBFS contents above 55%, there was almost no expansion at 30 days. The large decrease in ASR expansion from around 0.1% to less than 0.03% when the slag content is increased from 45% to 55% was explained by the double layer theory. Six different mortar mixtures were studied by energy dispersive x-ray analysis to determine the chemical composition of the alkali–silica gel produced. According to the double-layer theory, larger expansions are produced when monovalent ions (Na^+ and K^+) are more than the divalent ions (Ca^{2+} and Mg^{2+}) in the diffuse layer. The charge ratio, E_{BIV} that can be calculated by Equation 11.37 was found to be fairly correlated with the expansions measured as shown in Figure 11.15.

$$E_{BIV} = \frac{molCaO + molMgO}{molCaO + molMgO + molNa_2O + molK_2O} \qquad (11.37)$$

where, mol refers to the number of moles of the oxides.

(a)

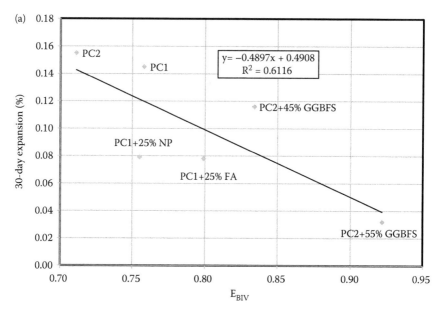

(b)

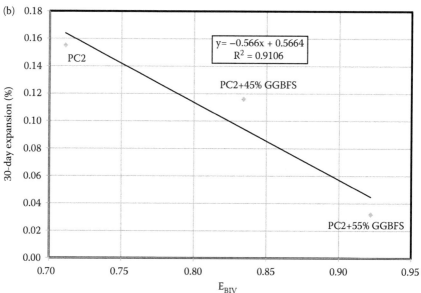

Figure 11.15 ASR expansion of mortar bars as a function of charge ratio, E_{BIV}. (a) Mixes with natural pozzolan, FA and GGBFS and (b) Mixes with GGBFS, only. (Data from Monteiro, P.J.M. et al. 1997. *Cement and Concrete Research*, 27(12), 1899–1909.)

Studies on trass (Andiç Çakır, 2007), Santorin earth (Mehta, 1981), pumicite (Cook, 1986), perlite (Bektaş et al., 2005), MK (Ramlochan et al., 2000) and zeolite (Naiqian et al., 1998) had all shown that these pozzolanic materials reduce the expansion caused by ASR. The effectiveness of pozzolans increases with the increase in amount and fineness.

Low-lime fly ashes were found to show similar effects in improving the resistance to alkali–silica expansion. Stark (1978), Nixon and Gaze (1983), Nixon et al. (1986) and Stark et al. (1993) have determined that low-lime fly ashes used to replace 15%–30% PC resulted in significant reductions in ASR expansions. However, using 5%–10% FA may increase the expansions when compared with non-FA concretes (Dunstan, 1981).

It is generally accepted that FA prevents ASR expansion if it is appropriately used. However, most of the research done so far for this purpose was on low-lime fly ashes (Shehata and Thomas, 2000). There are several reports that tried to correlate the effectiveness of fly ashes in mitigating ASR to their chemical compositions. In an extensive research conducted by Shehata and Thomas (2000) 18 different fly ashes with CaO contents varying from 5.57% to 30.00% and Na_2O_{eq} contents varying from 0.56% to 8.73% were used. Upon accelerated mortar bar testing, they have concluded that all the fly ashes used reduced ASR expansion and reduction becomes more pronounced with the increasing amount of FA. However, for a specific level of FA replacement, the expansions were generally increased with an increase in lime or alkali content or decrease in silica content of the FA which means that minimum amount of FA required to control the expansion becomes higher with an increase in lime or alkali content or decrease in silica content.

In another comprehensive investigation (Malvar and Lenke, 2006) the experimental results from five different previous researches were analysed. Thirty-one different fly ashes and five different PCs were used in the tests carried out according to ASTM C 1260. Fourteen of the fly ashes were Class F, sixteen were Class C, and one was neither Class F nor Class C. Two of the PCs were high-alkali, the other two were low-alkali and one was medium-alkali. In order to compare the results, the expansion of FA-incorporated cements was normalised by the expansion of their control cements. The linear correlations between various individual oxide contents of the blended cements and normalised expansions were determined first. Then, several combinations of oxides were defined as CaO_{eq} and SiO_{2eq}, as given by Equations 11.38 and 11.39 respectively, and they were related to ASR expansion. Finally a chemical index for blended cements, C_b, given by Equation 11.40 which resulted in maximised coefficient of determination was also related to ASR expansion.

$$CaO_{eq} = CaO + 0.905Na_2O + 0.595K_2O + 1.391MgO + 0.700SO_3$$
$$(11.38)$$

$$SiO_{2eq} = SiO_2 + 0.589Al_2O_3 + 0.376Fe_2O_3 \qquad (11.39)$$

The equivalencies in Equations 11.38 and 11.39 are obtained by considering molar ratios of alkalis, MgO, and SO_3 to CaO and Al_2O_3 and Fe_2O_3 to SiO_2, respectively.

$$C_b = \frac{CaO_{eq\alpha b}}{SiO_{2eq\beta b}} = \frac{CaO + \alpha(0.905Na_2O + 0.595K_2O + 1.391MgO + 0.700SO_3)}{SiO_2 + \beta(0.589Al_2O_3 + 0.376Fe_2O_3)}$$

(11.40)

where, $\alpha = 5.64$ and $\beta = 1.14$ are factors that maximise the coefficient of determination.

The results of the statistical analyses are summarised in Table 11.6.

At the end of extensive statistical manipulations Malvar and Lenke (2006) proposed an equation that would predict the minimum amount (W_{FA}) of a FA with a known chemical composition that would yield ASR expansion of 0.08% as specified by ASTM C 1260:

$$W_{FA} = \frac{1 - a_4 \tanh^{-1}\left(\dfrac{2\left(\dfrac{0.08}{\epsilon_{14C}}\right) - (a_1 + a_2)}{(a_2 - a_1)} \right) + a_3}{\left(1 - \dfrac{CaO_{eq\alpha FA}}{CaO_{eq\alpha C}}\right) - \left(1 - \dfrac{SiO_{2eq\beta FA}}{SiO_{2eq\beta C}}\right)\left(a_4\tanh^{-1}\left(\dfrac{2\left(\dfrac{0.08}{\epsilon_{14C}}\right) - (a_1 + a_2)}{(a_2 - a_1)} \right) + a_3 \right)}$$

(11.41)

where, a_1, a_2, a_3, a_4, α and β are 0, 1.0244, 0.6696, 0.1778, 6.0 and 1.0, respectively. ε_{14C}: 14-day ASR expansion of the PC used.

Similar results were obtained in another study in which six different fly ashes were used. The fly ashes were used either in their original form as obtained from the thermal power plant or they were reduced in size by sieving or grinding. The ASR expansion of the original fly ashes and fly ashes with average particle size greater than 10 μm was well-correlated with CaO, SiO_2, $SiO_2 + Al_2O_3 + Fe_2O_3$, $CaO + MgO + SO_3$, SiO_{2eq} and CaO_{eq} by exponential relations. For finer fly ashes (average particle size less than 10 μm) ASR expansions were linearly related to CaO, $CaO + MgO + SO_3$ and CaO_{eq} contents and logarithmically related to SiO_2, $SiO_2 + Al_2O_3 + Fe_2O_3$ and SiO_{2eq} contents of the fly ashes. All fly ashes having average particle sizes less than 5 μm were found to result in less than 0.10% ASR expansion limit, indicating that very fine fly ashes are very effective in mitigating ASR (Venkatanarayanan and Rangaraju, 2013).

The model proposed by Malvar and Lenke was further studied by Wright et al. (2014) on concretes containing recycled glass aggregates and

Table 11.6 Effect of different chemical parameters of FA-incorporated cements on normalised 14-day ASR expansion measure according to ASTM C 1260

Chemical parameter of the blended cement	Effect on normalised ASR expansion	Coefficient of determination, R^2
CaO	Increasing	0.7143
Alkalis (Na_2O_{eq})	No noticeable effect	–
MgO	Increasing	0.0552
SO_3	Increasing	0.5000
SiO_2	Decreasing	0.7409
Al_2O_3	Decreasing	0.6038
Fe_2O_3	Decreasing	0.1313
CaO_{eq}	Increasing	0.7799
SiO_{2eq}	Decreasing	0.7834
CaO_{eq}/SiO_{2eq}	Increasing	0.8287
C_b	Increasing	0.8671

Source: Adapted from Malvar, L.J., Lenke, L.R. 2006. *ACI Materials Journal*, 103(5), 319–326.

was found to be conservative. They proposed that the constants a_1 and a_4 should be revised. However, it should be noted that this later study was a laboratory investigation where ASR was generated using recycled glass sand which is seldom (if ever) encountered in real-life concretes.

GGBFS is another mineral admixture that is proven to reduce ASR expansion. Although some of the slags may have rather high amounts of total alkalis, maximum 50% of the total alkali content may be effective in ASR and the rest is bound in the glassy phase (Hobbs, 1988; Andiç-Çakır, 2007). In an experimental study carried out according to ASTM C 441 method in which pyrex glass was used as the reactive aggregate, GGBFS was found to reduce the ASR expansion significantly. The effectiveness of GGBFS increased with the increase in amount of GGBFS (Buck, 1987) as illustrated in Figure 11.16.

Thomas and Innis (1998) determined that 50% (by mass) GGBFS reduced the ASR expansion to less than 0.04% in concrete prisms prepared using a limestone aggregate containing reactive silica. In the same study, 35% GGBFS was found to be sufficient for concretes made from greywacke aggregate.

Kwon (2005) studied the AAR-inhibiting effect of GGBFS in high-strength concrete and determined that 30% replacement of PC by GGBFS resulted in significant reduction in the total alkali content of the concrete.

SF is also an effective mineral admixture in improving the ASR resistance of concrete. The amount of SF for controlling the expansion depends on the amount of available alkalis in the concrete. Generally, 5%–12% SF is considered to be sufficient (Thomas, 2013).

Concrete cores taken from existing structures in Iceland were studied to determine that even 5% (by mass of cement) SF in concretes resulted in the formation of much less alkali–silica gel formation when compared with non-SF concretes (Gudmundsson and Olafsson, 1998). Durand et al.

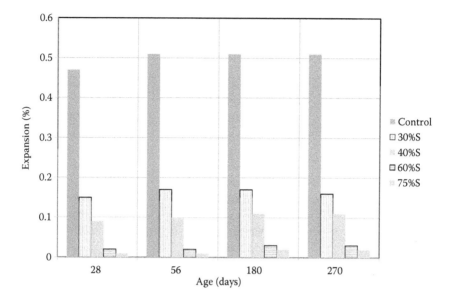

Figure 11.16 Effect of GGBFS on reducing ASR Expansion. (Data from Buck, A.D. 1987. Use of cementitious materials other than portland cement, in concrete durability, *Proceedings Katherine and Bryant Mather International Conference* [Ed. J.M. Scanlon], ACI SP-100, pp. 1863–1881.)

(1987) have used 5%, 10% and 15% (by mass) SF to partially replace the PC in concretes with argillite-containing reactive aggregates and found out that the expansions were reduced by 4%, 68% and 83%, respectively.

When used in ternary blends with GGBFS or FA, SF is efficient in controlling ASR expansions even at lower amounts around 4%–6% (Thomas, 2013).

Unlike the mitigation of ASR, mineral admixtures are not effective in reducing the expansions caused by ACR. However, they may have a beneficial effect through reducing the porosity of concrete and thus slowing the rate of pore liquid migration (ACI 221, 1998).

11.7 PERMEABILITY OF CONCRETE

Many of the concrete deterioration processes involve the ingress of moisture, certain aggressive liquids and gases into concrete as was discussed in the foregoing sections. Therefore, it is obvious that the permeability of concrete is of prime importance from the durability point of view. The lower the permeability of concrete the more durable it will be.

Concrete permeability depends on numerous factors such as (1) cementitious material content and type, (2) aggregate gradation, (3) compaction and (4) type and duration of curing. Besides all these, a factor that has the largest influence on durability is the water–cement (or water–cementitious)

ratio. As the water–cement ratio decreases, for a given degree of hydration, porosity of the cement paste decreases and the concrete becomes less
permeable. Capillary porosity, P_c, in a cement paste, total volume of hydration products, V_g, and the volume of unhydrated cement, V_u, are given by
Equations 11.42, 11.43 and 11.44, respectively (Mindess and Young, 1981):

$$P_c = \frac{w}{c} - 0.36\alpha \qquad\qquad\qquad (11.42)$$

$$V_g = 0.68\alpha \qquad\qquad\qquad (11.43)$$

$$V_u = (1 - \alpha)v_c \qquad\qquad\qquad (11.44)$$

where α is the degree of hydration and v_c is the specific volume of cement
(approximately equal to 0.32 for PCs).

The changes in the capillary porosity and volumes gel and unhydrated
cement with changing water–cement ratio for full hydration and with
changing degree of hydration for a given water–cement ratio of 0.50 are
shown in Figure 11.17.

The permeability of concrete depends mostly on the porosity of the
cement paste but it is also affected by the internal microcracks especially at
the aggregate–matrix interface. The flow of water through concrete can be
explained by D'Arcy's law for flow through porous media:

$$\frac{Q}{A} = K\frac{\Delta h}{l} \qquad\qquad\qquad (11.45)$$

where, Q is the rate of flow, A the cross sectional area of the test specimen,
Δh head of water, l the thickness of the test specimen and K the coefficient
of water permeability. K is not constant. It is strongly dependent on the
porosity of the cement paste, therefore, it changes with the water–cement
ratio and degree of hydration.

The water permeability test of concrete is carried out on samples with
sealed side surfaces and by applying pressurised water from one face and
measuring the flow from the opposite face. There may be several problems
encountered in this test such as the difficulty of attaining a steady-state
flow, possible leakage of water from the side surfaces, different permeability characteristics of the surface layer and the core, etc. Furthermore, the
test may take a long period of time, weeks or even months, to achieve
a measurable flow for high-performance concretes (Bentz et al., 1999).
Therefore, it is more common to use other shorter term test methods such
as measuring depth of water penetration under pressure (TS EN 12390-8,
2010), measurement of rate of absorption (ASTM C 1585, 2004) and rapid
chloride permeability (ASTM C 1202, 2004) in determining the permeability characteristics of concrete.

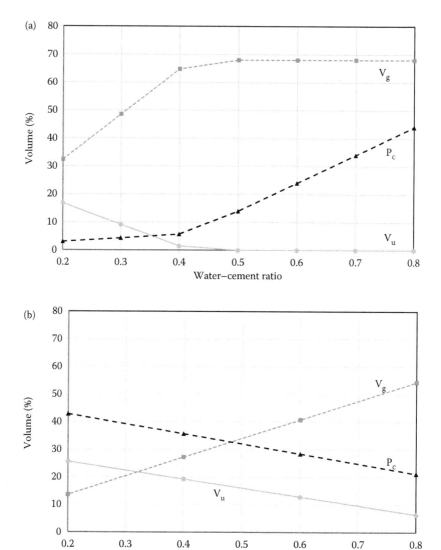

Figure 11.17 Volumes of gel, unhydrated cement and capillary porosity in a portland cement paste (a) upon full hydration for different water–cement ratios and (b) at different degrees of hydration with a constant water–cement ratio of 0.50. Note that full hydration can be achieved at water–cement ratio at or above 0.42.

One of the first reports on the effects of mineral admixtures on concrete permeability was that of Davies (1954) in which 30% and 60% FA were used to replace the PC in the manufacturing of concrete pipes. Although the early values were higher, FA incorporation was found to result in more than 70% reduction in the permeability.

Hooton (1993) carried out permeability tests on SF-incorporated cement pastes and reported that after 7 days the coefficient of permeability reduces significantly.

In a recent study on the effect of various mineral admixtures on the permeability properties of self-consolidating concretes, the permeability characteristics of 17 SCCs containing SF, MK, low-lime fly ash, high-lime fly ash and GGBFS were investigated through chlorine ion permeability, water penetration depth, water absorption and sorptivity (Ahari et al., 2015). Binary blends of 4%, 8% and 12% SF, 18% and 36% low-lime and high-lime fly ashes, 8%, 18% and 36% MK and 18% GGBFS; ternary blends containing 18% GGBFS and 8% SF, 36% low-lime and high-lime fly ashes, and 36% MK; and a quaternary blend of 8% SF, 18% high-lime fly ash and 18% GGBFS were used as cementitious materials besides the control PC. All mineral admixtures were used to partially replace the PC, by mass. The water–cementitious ratio and workability of all mixes were kept constant by adjusting the superplasticiser amount. The results obtained are given in Figure 11.18. As can be seen from the figure, mineral admixture incorporation resulted in reduced chloride ion permeability at both test ages. The positive effect was more pronounced at 90 days.

Considering the limits given in ASTM C 1202 (2004), all mineral admixture containing mixtures may be classified as having low (1000–2000 C) or very low (100–1000 C) permeability. The measure of the ease of water flow under pressure through the concrete was obtained as water penetration depth. Although the low-lime fly ash mixtures had water penetration depths close to those of the control, all mixtures had lower penetration depths than the control mixture, at both ages. Similar trends were observed for water absorption of immersed specimens determined according to ASTM C 642 (2004) and sorptivity indices determined according to ASTM C 1585 (2004). However, low-lime fly ash specimens had higher water absorption and sorptivity index than those of the control, at 28 days.

Valipour et al. (2013) used natural zeolite, SF and MK to partially replace the PC and determined that all three mineral admixtures improved the water absorption, sorptivity, electrical resistance, gas permeability and chloride ion diffusion of concretes significantly with respect to the control concrete without any mineral admixture.

In another study on the permeability characteristics of concretes, Tsivilis et al. (2003) compared six portland limestone cements with a control PC. They prepared two series of concretes. The first series consisted of four mixtures containing 0%, 10%, 15% and 20% limestone powder and the second series consisted of three mixtures containing 20%, 25% and 35% limestone powder. The water–cementitious ratios of the first and second series were 0.70 and 0.62, respectively. The cementitious material contents were 270 and 330 kg/m³ of concrete, respectively. The specific surface areas of the cements ranged from 260 to 530 m²/g with the control cement being the coarsest. Gas permeability, water permeability and sorptivity of the

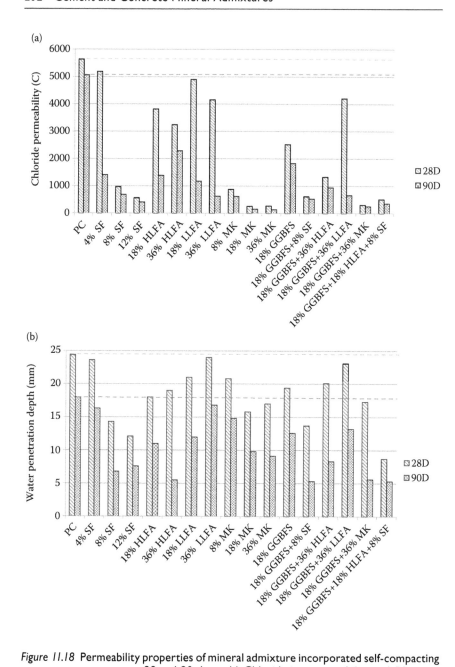

Figure 11.18 Permeability properties of mineral admixture incorporated self-compacting concretes at 28 and 90 days: (a) Chloride ion permeability and (b) water penetration depth. (Data from Ahari, R.S., Erdem, T.K., Ramyar, K. 2015. *Construction and Building Materials*, 79, 326–336.)   (*Continued*)

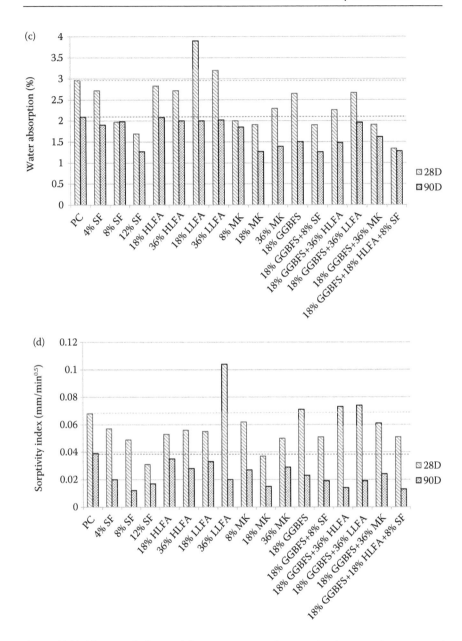

Figure 11.18 (Continued) Permeability properties of mineral admixture incorporated self-compacting concretes at 28 and 90 days: (c) Water absorption and (d) sorptivity index (Data from Ahari, R.S., Erdem, T.K., Ramyar, K. 2015. *Construction and Building Materials*, 79, 326–336.)

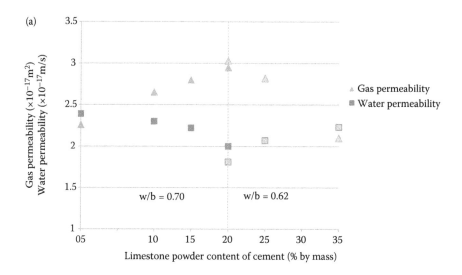

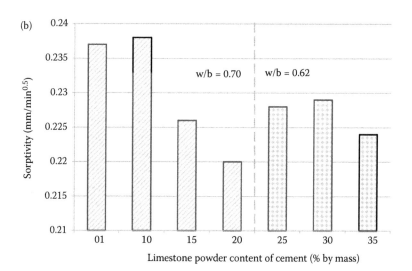

Figure 11.19 (a) Gas and water permeability and (b) sorptivity of portland limestone concretes at 28 days. (Data from Tsivilis, S. et al., *2003. Cement and Concrete Research*, 33, 1465–1471.)

seven concretes were determined at 28 days. All concretes had similar consistencies and unit weights in fresh state. It was determined that concretes made with portland limestone cements had slightly higher gas permeability coefficients and slightly lower water permeability coefficients and sorptivity indices than the control PC concretes, as shown in Figure 11.19.

Chapter 12

Proportioning mineral admixture-incorporated concretes

Concrete mix proportioning, which is also referred to as concrete mix design, is a process that consists of two basic parts that are interrelated with each other: (1) selection of suitable concrete ingredients namely, cement, aggregates, water, mineral and chemical admixtures and (2) determining their relative proportions with due care to maintaining a balance among (a) economy, (b) workability, (c) strength, (d) durability, (e) density and (f) appearance of the resulting concrete.

The cost of concrete consists of the costs of ingredients (material costs), labour and the equipment. Labour and equipment costs are more or less the same unless the concrete is a special one. Therefore, material costs are usually the determining ones in making a concrete mix design. Considering the costs of the ingredients per unit volume of concrete produced, it is generally the cement content that governs the cost of a concrete. In order to keep the cement content at a minimum possible amount, lower consistencies that will not impair the placeability and uniformity of concrete, increased maximum aggregate size, higher coarse aggregate-to-fine aggregate ratios, mineral admixtures, and where necessary, chemical admixtures may be used. Besides all these, good quality control is another important factor in reducing the cost of concrete (Mindess and Young, 1981).

Workability, as discussed in Chapter 9, is that property of fresh concrete which determines its capacity to be placed, consolidated, and finished properly with minimum segregation and bleeding. It embodies both cohesiveness and mobility and is affected by the proportions, grading and particle shape of aggregates, amount and quality of cementitious materials, presence of chemical admixtures, air-entrainment and the consistency of the mix. The consistency of properly proportioned concrete depends mainly on the water content.

Generally, strength is considered as the most important characteristic of concrete and most specifications require a minimum compressive strength. As discussed in Chapter 10, strength is affected by numerous factors which can be categorised into four major groups as properties and relative proportions of the materials constituting concrete, preparation and curing methods and test conditions. For a given set of materials, methods and conditions, concrete strength is determined by water–cementitious ratio. Although 28-day strength is frequently used for mix proportioning, structural design and evaluation of concrete; strength at other ages may also be considered.

Concrete must be able to resist deteriorative environmental exposures that may reduce or completely end its serviceability. Details of concrete durability were discussed in Chapter 11. Although there are many factors affecting the durability characteristics of concrete, permeability which is directly related with water–cementitious ratio is of paramount significance.

Density of any solid material is inversely proportional to its porosity. Therefore, density of concrete may also be an important consideration in concrete mix design. Besides, although the density of normal concrete is around 2,400 kg/m³, concretes with densities as low as 400 kg/m³ and as high as 5,500 kg/m³ may be needed for some special purposes such as insulation and radiation shielding, respectively. In such cases, the concrete mix proportioning differs from that of normal concretes.

Although there are still some people, even in the engineering environment, who think that concrete mix proportioning is easy and several simple recipes would be sufficient for many concreting jobs, it is not. Concrete mix proportioning is a process that involves numerous variables and dimensions. The variety and range of concrete components and properties within the whole life cycle and sustainability issues that necessitate a balance between environmental demands and technical requirements make concrete mix design a rather complicated process.

12.1 PRINCIPLES OF CONCRETE MIX PROPORTIONING

Before starting the mix design calculations, the characteristics of the concrete must be selected properly, based on its intended use. These characteristics should reflect the requirements for the concrete both in fresh and hardened states. After selecting the concrete characteristics, mix proportioning can be done on the basis of test data or field experience of the concrete ingredients actually to be used. If such information is not available, the estimates given in standards or recommendations related with concrete mix proportioning may be employed. In any case, preparation of a trial mix to check whether some (or sometimes all) of the characteristics required are achieved or not. In other words, concrete mix proportioning is usually an iterative process by which necessary adjustments are made for the following trials until the requirements are achieved.

Information available on (a) sieve analyses of the aggregates, (b) unit weight of coarse aggregate, (c) fineness modulus of fine aggregate, (d) bulk specific gravities and absorption of aggregates, (e) mixing water requirement, (f) relationship between strength and water–cementitious ratio, and

(g) densities of cements and mineral admixtures would be useful in concrete mix proportioning (ACI 211, 1997).

There are different concrete mix design procedures used in different countries. Some of them are standardised while some are recommendations. However, all of them serve the same purpose of proper proportioning of the ingredients once the workability, strength and durability characteristics of the concrete are selected. Determination of the relative proportions of the ingredients for normal concrete is usually accomplished in the following sequence:

1. *Choice of maximum aggregate size, D_{max}*: The voids content of continuously graded aggregates with larger nominal maximum size is less than that of smaller maximum sizes. Therefore, shifting the grading curve of the aggregate to the coarser side results in less mortar requirement per unit volume of concrete. Generally, the maximum size of the aggregate should be the largest as long as it is consistent with the dimensions of the structure. Usually, the structural limitations require D_{max} to be smaller than or equal to 1/5 of the narrowest dimension between the sides of the forms, 1/3 of the depth of the slabs, or 3/4 of the clear spacing between the reinforcement bars. On the other hand, even if the structural limitations may allow the use of D_{max} larger than ~40 mm, it is not recommended for normal concretes. Two counteracting parameters are affected by increasing the maximum aggregate size of a well-graded aggregate: Decreased specific surface area of the aggregate results in decrease in (1) water–cement ratio and (2) bond between the coarse aggregate and the matrix. The former dominates upto maximum aggregate sizes around 40 mm. Beyond that size, the latter becomes more effective and may cause strength reduction.

2. *Estimation of mixing water content*: For a well-graded aggregate, mixing water requirement of a concrete with a specified consistency depends basically on the maximum aggregate size. Recommendations or standards on concrete mix proportioning usually provide tables or charts that relate mixing water content to maximum aggregate size, for a specified slump value. The tables also give estimates of entrapped air contents for non-air-entrained concretes and recommend average air contents for air-entrained concretes. Aggregate shape, using AEA or chemical admixtures would also affect the water requirement. Angular aggregates may require more water than rounded aggregates. However, their bond strength is higher than that of rounded ones. Therefore, the strengths of normal concrete made from rounded or angular aggregates are similar as long as the aggregates have similar particle-size distributions. Entrained air has a lubricating effect in fresh concrete that results in lower water requirement for a given consistency. Water-reducing or high-range water-reducing admixtures provide better workability and therefore water content may be

reduced when they are used. However, care should be taken that they should not cause excessive segregation or bleeding.

3. *Selection of water–cementitious ratio:* Water–cement or water–cementitious ratio is the prime factor affecting the strength and durability of a concrete. Different cements, mineral admixtures and aggregates may result in different strengths and durability characteristics at the same water–cementitious ratio. Therefore, if data on the relationship between water–cementitious ratio and strength and durability characteristics are available, it is strongly recommended to use them. If not, tables given in the standards or recommendations for concrete mix proportioning may be used as a first estimate. Such tables provide minimum and maximum water–cementitious ratios for a given strength or a given environmental exposure, respectively.

4. *Calculation of cement content:* Cement content of the concrete is calculated by dividing the water content obtained in step 2 by the water–cement ratio obtained in step 3. If the job specification requires a minimum cement content, the larger of the two must be used and adjustment in the water content should be made according to the water–cement ratio selected in step 3.

5. *Calculation of aggregate content:* Mixing water, air and cement (and mineral admixture) contents determined previously are for unit volume of concrete. Therefore, total volume of aggregates, V_{ta} can be calculated as

$$V_{ta} = 1 - V_c - V_{ma} - V_w - V_a - V_{cha} \qquad (12.1)$$

where, V_c, V_{ma}, V_w, V_a and V_{cha} are volumes of cement, mineral admixture(s), water, air and chemical admixture(s) used.

Mineral admixtures may be incorporated in concrete as a part of blended cement or separately. If used as part of blended cement V_{ma} need not be calculated. Chemical admixtures are used in concrete in small amounts described as mass percentage of cement. So, to convert masses of the ingredients into volumes in Equation 12.1, their densities must be known.

There are two slightly different approaches for calculating the aggregate content. When aggregate is used in two size groups as coarse and fine aggregates, tables that relate the fineness modulus of fine aggregate and maximum aggregate size to the volume of oven-dry-rodded coarse aggregate per unit volume of concrete are given in the recommendations for concrete mix proportioning, as in ACI 211 (1997). Then the oven-dry mass of coarse aggregate is calculated by multiplying the value obtained by the dry-rodded unit weight of it.

Finally, as the masses (or volumes) of all ingredients of concrete except the fine aggregate are determined, fine aggregate content may be determined volumetrically or on mass basis by Equations 12.2 or 12.3, respectively.

$$V_{fag} = 1 - V_c - V_{ma} - V_w - V_{cag} - V_{cha} \qquad (12.2)$$

where V_{fag} and V_{cag} are the volumes of fine and coarse aggregates, respectively. Coarse aggregate volume is calculated by dividing its mass by the density.

$$M_{fag} = M_{conc} - M_c - M_{ma} - M_w - M_{cag} - M_{cha} \qquad (12.3)$$

where, M stands for mass and M_{conc} is the mass of fresh concrete per unit volume. Estimates for mass of fresh concrete per unit volume (kg/m³) are given in ACI 211 for different maximum aggregate sizes.

If, instead of using an estimate for the unit mass of fresh concrete a more precise value is preferred, Equation 12.4 may be used.

$$M_{conc} = 10\bar{G}_{ag}(100 - A) + M_c\left(1 - \frac{\bar{G}_{ag}}{G_c}\right) - M_w(\bar{G}_{ag} - 1) \qquad (12.4)$$

where, $\bar{G}_{ag}$ is weighted average bulk specific gravity of fine and coarse aggregates in saturated, surface-dry condition (SSD); G_c is the specific gravity of cement; and A is the air content (%).

When several different size groups of aggregates are to be combined, first the relative mass proportions of each size group should be determined according to the particle-size distribution curves recommended in the standards for different maximum aggregate sizes. The combined aggregate should have a particle-size distribution curve that fits into the limits given in the standard. An example for particle-size distribution limits for an aggregate with $D_{max} = 31.5$ mm is given in Figure 12.1. Total aggregate volume is calculated by Equation 12.1 and mass of each individual size group can be calculated by

$$M_{agg,i} = V_{ta}\bar{\rho}_{ta}X_i \qquad (12.5)$$

where $M_{agg,i}$ is the mass of size group i, X_i the relative size proportion of it in the total aggregate mass, and $\bar{\rho}_{ta}$ the weighted average density of the aggregate mixture.

6. *Aggregate moisture adjustment:* There are four possible moisture conditions of aggregates: (1) oven-dry (OD), (2) air-dry (D), (3) SSD and (4) wet (W). OD and SSD are the moisture conditions that can be attained under laboratory conditions. There is no moisture in OD

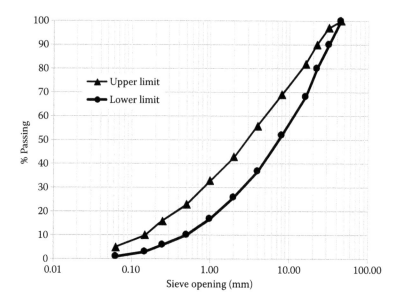

Figure 12.1 Particle size distribution limits for an aggregate mix with maximum aggregate size of 31.5 mm which can be used in a concrete with sufficient pumpability. (Adapted from TS 802. 2005. Design Criteria for Concrete Mixtures. Turkish Standards Institute, Ankara.)

aggregate. All permeable pores are completely filled with moisture but the surface of the aggregate is dry in SSD condition. W or D are field conditions.

All moisture conditions of an aggregate are given as percentages of its OD weight and mixing water estimation assumes that the aggregate is in SSD condition. So, both of these and the actual moisture condition in the field should be determined in order to make the necessary moisture adjustments. If the aggregate is wet, excess moisture should be reduced from the net mixing water content estimated or if the aggregate is air dry, the amount of water that it will absorb should be added to the net mixing water content estimated in step 3.

7. *Trial batch adjustments:* Usually a trial batch (~0.20 m³) is prepared with the calculated relative proportions of the ingredients to check if the slump, unit weight, air content and yield of concrete are appropriate.

If the slump of the trial batch is not correct, mixing water content will be increased or decreased by approximately 2 kg/m³ of concrete for each 10 mm difference from the required value. The changed mixing water content requires the cement content to be adjusted for a specified water–cement ratio. Thus, recalculation starting from step 4 becomes necessary.

If the mix proportioning was made on mass basis and the unit weight of trial batch concrete differs from the estimated value significantly, recalculations with the new unit weight should be made.

When the air content of the air-entrained concrete is not correct, the air-entraining agent content should be reestimated and the mixing water content will be increased or decreased by approximately 3 kg/m^3 of concrete for each 1% difference of measured air content from the desired value.

12.2 MIX PROPORTIONING CONCRETE WITH MINERAL ADMIXTURES

In order to achieve an efficient utilization of mineral admixtures in concrete, their characteristics in relation to and in combination with the other concrete ingredients have to be considered. The extent of such considerations mainly depends on how a mineral admixture is introduced into the concrete mixture. There are two basic approaches for the inclusion of mineral admixture into concrete: (1) using blended cements containing the mineral admixture(s) or (2) introducing mineral admixtures as a separate ingredient of concrete.

The first approach is simpler than the second one since common mix proportioning methods used for PC concretes can be employed for concretes to be made with many different types of blended cements containing mineral admixtures that comply with the relevant standards. Mineral admixture incorporation in such cements is done either by intergrinding or separate grinding and then blending. An obvious advantage of blended cements is that they are an intimate mixture of PC and mineral admixtures thus result in a more homogeneous product. On the other hand, the user will be restricted to only the available PC–mineral admixture ratios.

The second approach permits the use of PC–mineral admixture ratios that would meet the property requirements of the resulting concrete better and may be more economical. Basically, mix proportioning of concretes containing mineral admixtures as a separate ingredient may be done in two ways: (1) simple partial replacement of PC by mineral admixture or (2) partial replacement and addition, by taking the influence of mineral admixtures on various required properties of the concrete into account.

12.2.1 Simple partial replacement of PC

The simplest way of incorporating mineral admixtures is direct replacement of a part of PC, usually on a mass basis. The total binder volume in the mixture increases since the density of the mineral admixture used is lower than that of the PC. This would require the adjustment of the aggregate and water contents, depending on the extent of changes it causes in the workability of the concrete mixture. An important problem involved with this method is the possible reduction in early-age strength. When high proportions of mineral admixtures are used, generally strengths equivalent with the corresponding PC concrete may be reached at later ages when latent hydraulic or pozzolanic admixtures are used or may never be reached in the case of less reactive mineral admixtures. However, this depends on the type, nature and amount of the admixture and the PC used.

12.2.2 Partial replacement and addition

To overcome the lower early strength problem that may be encountered in the simple replacement method, instead of using the same amount as the PC removed, a greater amount of mineral admixture is put in the concrete mixture, by considering the properties of the mineral admixture such as particle size, chemical composition and mineralogy that may affect the workability and strength of the concrete. Of course, the amount of PC removed, type and quality of PC, presence of any chemical admixtures in the mix and the conditions of placing and curing should also be taken into account.

The results of an experimental study (Ramyar, 1993) that compares the two ways of incorporating fly ashes into concrete are summarised below. The concrete mixtures were designed to have similar slumps (50–75 mm). Two high-lime (HL1 and HL2) and a low-lime (LL) fly ashes were used. 10%, 20% and 40% (by mass) PC were removed from the control concrete mixture and (1) the same amounts of FA was used by direct replacement and (2) twice and four times more (by mass) FA amounts were used for 10% PC removal and twice more FA amount was used for 20% PC removal by partial replacement and addition. Fine aggregate contents of the mixtures were adjusted in order to keep the slump value between 50 and 75 mm. Figures 12.2 and 12.3 exemplify the changes in water-binder ratio and strength of the concretes for a specified slump, upon FA incorporation.

12.3 WATER–CEMENT AND WATER–CEMENTITIOUS RATIO AND EFFICIENCY FACTOR CONCEPTS

When mineral admixtures are used in concrete, water–cementitious ratio must be considered instead of the traditional water–cement ratio due to the differences in the specific gravities of PC and mineral admixtures, the latter being lighter than the former, normally. Besides, the efficiency of a mineral admixture used in concrete may change depending on many parameters related with the mineral admixture and the cement as well as the mix proportions, curing conditions and age. Therefore, in designing mineral admixture-incorporated concrete mixtures these two points should also be considered.

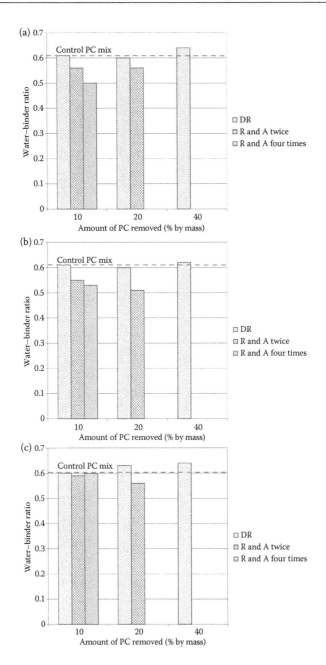

Figure 12.2 Changes in water–binder ratios of (a) LL, (b) HL1 and (c) HL2 fly ash concretes for a specified slump with respect to that of the control PC concrete upon direct replacement (DR) and replacement and addition (R and A). (Data from Ramyar, K. 1993. Effects of turkish fly ashes on the portland cement-fly ash systems, PhD thesis, Middle East Technical University, 208pp.)

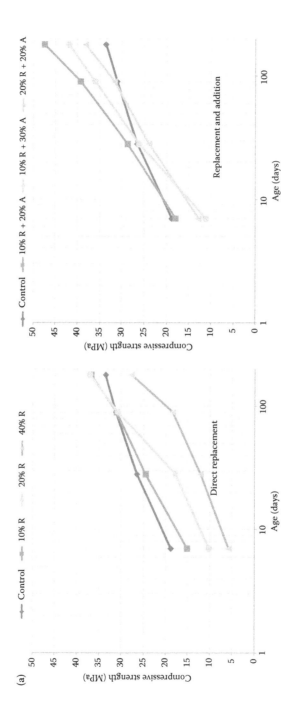

Figure 12.3 Changes in compressive strengths of (a) LL, fly ash concretes for a specified slump with respect to that of the control PC concrete upon direct replacement and replacement and addition. (Data from Ramyar, K. 1993. Effects of turkish fly ashes on the portland cement-fly ash systems, PhD thesis, Middle East Technical University, 208pp.) (Continued)

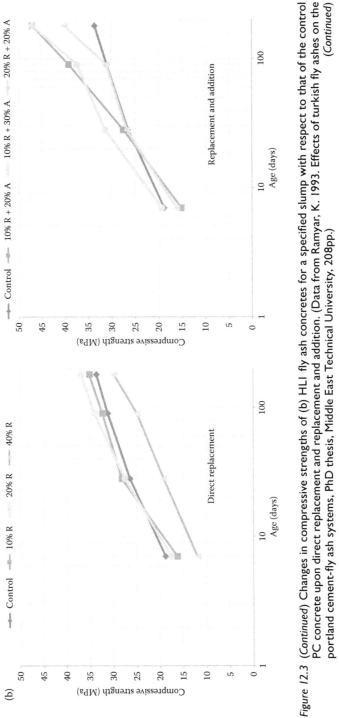

Figure 12.3 (Continued) Changes in compressive strengths of (b) HLI fly ash concretes for a specified slump with respect to that of the control PC concrete upon direct replacement and replacement and addition. (Data from Ramyar, K. 1993. Effects of turkish fly ashes on the portland cement–fly ash systems, PhD thesis, Middle East Technical University, 208pp.)

(Continued)

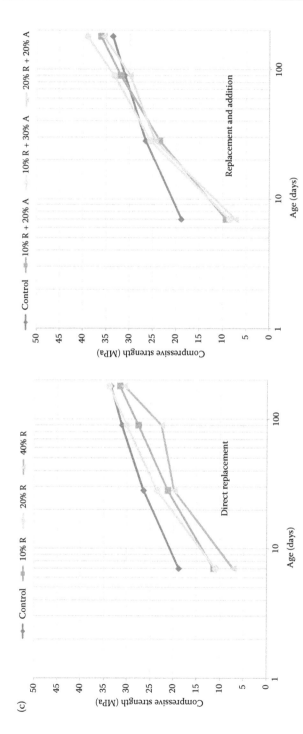

Figure 12.3 (Continued) Changes in compressive strengths of (c) HL2 fly ash concretes for a specified slump with respect to that of the control PC concrete upon direct replacement and replacement and addition. (Data from Ramyar, K. 1993. Effects of turkish fly ashes on the portland cement-fly ash systems, PhD thesis, Middle East Technical University, 208pp.)

12.3.1 Water–cementitious ratio

In concrete mix design, when blended cements are used, water–cementitious ratio is often considered to be synonymous to water–cement ratio; the former meaning the water-to-blended cement ratio and the latter meaning the water-to-PC ratio. However, when a mineral admixture is used as a separate ingredient in concrete, the water–cementitious ratio should be taken into account by considering either the mass or the volume equivalency of the water–cementitious ratio to the water–cement ratio of concrete made with only PC (ACI 211, 1997). The two approaches are described in ACI Committee 211 report with examples.

The amount of mineral admixture used may be expressed as either mass percentage (F_w) or volume percentage (F_v) of the total cementitious material.

$$F_w = \frac{MA}{PC + MA} \tag{12.6}$$

where, MA and PC are the masses of mineral admixture and PC, respectively, and F_w is the mineral admixture percentage factor by mass.

Mass and volume percentage factors are related with each other through the specific gravities of the PC and mineral admixture as follows:

$$F_v = \frac{1}{1 + (G_{MA}/G_{PC})((1/F_w) - 1)} \tag{12.7}$$

In other words, when mass-equivalency approach is taken,

$$\frac{W}{PC + MA} = \frac{W}{PC} \tag{12.8}$$

When volume-equivalency approach is taken,

$$\frac{W}{PC + MA} = \frac{G_{PC}(W/PC)}{G_{PC}(1 - F_v) + G_{MA}(F_v)} \tag{12.9}$$

Thus, the water–cementitious ratio determined on volume-equivalency basis becomes larger as the relative proportion of mineral admixture increases so that the same volumetric water–cementitious ratio will be maintained. Figure 12.4 compares water–cementitious ratios on the weight and volume-equivalency bases.

It is necessary to note that, for a specified water content of a concrete mixture, the volume-equivalency approach leads to lower total cementitious materials content. The total volume of PC and mineral admixture will be greater than that of PC in mass equivalency whereas total mass

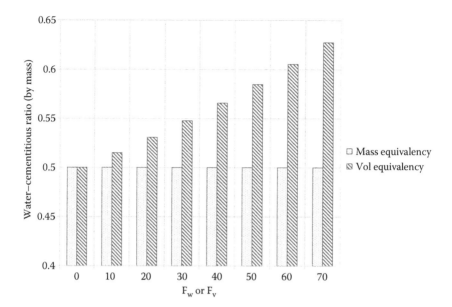

Figure 12.4 Water–cementitious ratios of PC–mineral admixture mixtures on weight-
and volume-equivalency bases. It is assumed that PC and mineral admixture
have specific gravities of 3.10 and 2.20, respectively.

of PC and mineral admixture will be smaller than that of PC in volume
equivalency. On the other hand, for a specified amount of total cementitious material, the PC content will be higher in volume equivalency than in
weight equivalency, as illustrated in Figure 12.5.

12.3.2 Efficiency factor

Efficiency of a mineral admixture in concrete may be broadly defined as
its relative proportion to PC that it replaces without changing the properties of the concrete. I. A. Smith defined a *fly ash cementing efficiency
factor*, K in his paper that proposed a method of mix proportioning for
FA-incorporated concretes (Smith, 1967). Actually, in order to explain K,
Smith first defined an *effective water–cement ratio*, $(W/C)_e$ which is the
water–cement ratio of the PC concrete that gives the same strength as the
FA-incorporated concrete. Thus, K is defined as the binding efficiency of
FA as measured by its effect on $(W/C)_e$. Simply, it is the mass of FA used
(M_{FA}) that is equivalent to a $K \cdot M_{FA}$ mass of PC. Based on the results of
an experimental investigation on British fly ashes, he proposed K as 0.25.

Later on, numerous investigations on determining the efficiency of different mineral admixtures in relation to strength and various durability
aspects were carried out. Babu and Rao (1996) evaluated the efficiency of

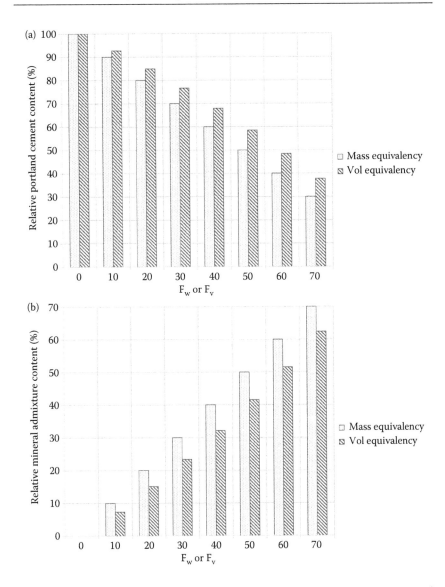

Figure 12.5 Relative proportions of (a) PC and (b) mineral admixture in a mixture with constant cementitious material content. It is assumed that PC and mineral admixture have specific gravities of 3.10 and 2.20, respectively.

fly ashes in by considering the water–cementitious ratio–strength relations, age and amount of FA replacing the cement. They defined three efficiency factors as (1) a 'general efficiency factor', k_e which is dependent on age, (2) a 'percentage efficiency factor', k_p which is dependent on the amount of FA replacing the cement and (3) an 'overall efficiency factor', k which is the sum of k_e and k_p.

Babu and Rao (1996) first tried to bring the strength values of FA concretes close to that of their control concretes made from PC using [W/(C + k_eFA)] and found out that the general efficiency factor, k_e, by itself, is not sufficient for all replacement levels. Then, by evaluating the strength difference through the percentage efficiency factor they were able to bring the strengths of FA concretes closer to those of control concretes using [W/(C + k_eFA + K_pFA)]. Finally, they proposed k_p and k for different ages as Equations 12.10 and 12.11, respectively.

$$k_p = 2.54p^2 - 3.62p + 1.13 \qquad (12.10)$$

where p is the mass ratio of FA to total cementitious material.

$$
\begin{aligned}
k_7 &= 2.67p^2 - 3.75p + 1.45 \\
k_{28} &= 2.78p^2 - 3.80p + 1.64 \qquad (12.11) \\
k_{90} &= 2.50p^2 - 3.59p + 1.73
\end{aligned}
$$

where subscripts 7, 28 and 90 indicate the age of concrete in days.

Similar reasoning was applied for GGBFS-incorporated concretes. The efficiency factors, k_e, k_p and k, at 28 days, were determined as 0.90, 0.39 to −0.20 and 1.29 to 0.70 for PC replacement levels of 10%–80% by GGBFS (Babu and Kumar, 2000).

Papadakis and Tsimas (2002) studied the efficiencies of two natural pozzolans, three high-lime fly ashes, and a nickel slag from strength, chlorine ion penetration, and carbonatation resistance points of view. Fly ashes were found to be equivalent to PC (k = 1.0) when strength is considered. On the other hand, natural pozzolans had k = 0.2–0.3 and nickel slag had almost no efficiency. When chlorine ion penetration resistance is considered, efficiency factors were determined as 2.0–2.5 for fly ashes and 1.0 for natural pozzolans. Efficiency under carbonatation was determined for only one of the fly ashes as 0.7.

Simultaneously with the previous study, Papadakis et al. (2002) proposed the following empirical relationship to estimate the strength efficiency factor for fly ashes:

$$k = \left(\frac{\gamma_s f_{S,P}}{f_{S,C}} \right)\left(1 - \frac{aW}{C} \right) \qquad (12.12)$$

where

$f_{S,P}$ and $f_{S,C}$ are the mass fractions of SiO_2 in FA and PC, respectively;
γ_s is the ratio of reactive silica (as determined by EN 196-2) to total silica in the FA;

a is a parameter that depends on age and was estimated as 1.06, 0.72, 0.50 and 0.23 for 2, 7, 28 and 90 days, respectively;

W and C are the water and cement contents.

It should be noted that, Equation 12.12 was proposed after studying only three high-lime fly ashes. Later, Antiohos et al. (2007) showed that it is applicable to three other similar high-lime fly ashes. Therefore, although it seems promising, it should be verified for other fly ashes, also.

An attempt to apply the efficiency factor concept to GGBFS-incorporated concretes with respect to durability issues such as acid attack, carbonatation, sulphate attack, chloride ingress and alkali–silica reaction concluded that the results are ambiguous in spite of the laborious work done (Gruyaert, et al., 2013). Aponte et al. (2012) studied the durability efficiency of fly ashes by means of chloride diffusion. They used two PCs, CEM I 42.5 and CEM I 52.5 and a high-lime and two low-lime fly ashes to replace 25% and 43% (by mass) of the PCs. Three series of concretes were prepared with water–cementitious ratios of 0.40, 0.525 and 0.65. Although they had obtained some numerical results, these were far from being conclusive, either.

The European standard on concrete, EN 206-1 (2014), specifies k for FA as 0.4 when used with CEM I cements for FA–PC ratios less than or equal to 0.33 (by mass) and when used with CEM II/A cements for FA–cement ratios less than or equal to 0.25. For SF, k is taken as 2.0 when it is used with CEM I and CEM II/A (except CEM II/A-D) cements as long as SF–cement ratio is less than or equal to 0.11 (by mass). However, when SF-incorporated concrete will be exposed to carbonatation or freezing-thawing, k should be taken as 1.0. EN 206-1 recommends k = 0.6 for GGBFS for slag–cement ratios less than or equal to 1.0. Although higher amounts of FA, SF or GGBFS than those specified in the standard are allowable, the excess amount should not be considered in water–cementitious ratio calculations given by

$$\text{Water} - \text{cementitious ratio} = \frac{\text{Water}}{\text{Portland cement} + k \cdot \text{mineral admixture}}.$$

$$(12.13)$$

Since there are so many factors related with the characteristics of the mineral admixtures such as chemical composition, mineralogy, pozzolanicity, particle shape, size and distribution and cement such as type, compound composition and fineness as well as the parameters of the concrete mix such as total cementitious material content, relative proportion of mineral admixture, water–cementitious ratio and curing method and duration that influence the efficiency of a mineral admixture, it seems that it is very difficult, if not impossible, to obtain universally applicable k-values and those given in standards or recommendations may only be used as a guide.

Chapter 13

International standards on mineral admixtures in cement and concrete

Standards for the use of mineral admixtures in cementitious systems may be grouped into two. The first group is related to the properties of mineral admixtures to be used as basic ingredients in cements, whereas the second group is related to their properties when used as a basic ingredient in concrete. Many countries throughout the world have their own standards on mineral admixtures for cement and concrete. However, the discussion here will be confined to the ones prepared by two widely known international standards organisations, the American Society for Testing and Materials (ASTM) and the European Committee for Standardization (CEN) since they are more commonly accepted and used. Standard specifications issued by ASTM and CEN on mineral admixtures for concrete and blended cements are listed in Table 13.1.

13.1 OVERVIEW OF THE CEMENT STANDARDS

The European cement standard, EN 197-1 (2012) specifies GBFS, natural and calcined pozzolan, siliceous and calcareous FA, SF, burnt shale and limestone powder as possible major constituents of four main types of cements. Besides, all these materials may also be used as minor additional constituents not exceeding 5% (by mass) in all five main types of cements.

ASTM C 595 (2014) is a specification for blended hydraulic cements containing BFS, pozzolan, limestone or combinations of these with PC. These cements do not contain other additional materials except air-entraining, processing and functional additives. Some slag-incorporated blended cements are permitted to contain hydrated lime.

13.1.1 Requirements for mineral admixtures in EN 197-1

European cement standard specifies five main types of cements as CEM I (PC), CEM II (portland-MA cement), CEM III (BFS cement), CEM IV (pozzolanic cement) and CEM V (composite cement). 'MA' stands for the

Table 13.1 ASTM and CEN standard specifications related with the use of mineral
admixtures in cement and concrete

ASTM	CEN
For use as an ingredient of cement	
ASTM C 595 Standard Specification for Blended Hydraulic Cements	EN 197-1 Cement. Composition, Specifications and Conformity Criteria for Common Cements
ASTM C 1157 Standard Performance Specification for Hydraulic Cement	
For use as a concrete component	
ASTM C 618 Standard Specification for Coal Fly Ash and Raw or Calcined Natural Pozzolan for Use in Concrete	EN 450-1 Fly Ash for Concrete. Definition, Specifications and Conformity Criteria
ASTM C 989 Standard Specification for Slag Cement for Use in Concrete and Mortars	EN 13263-1 Silica Fume for Concrete. Definitions, Requirements and Conformity Criteria
ASTM C 1240 Standard Specification for Silica Fume Used in Cementitious Mixtures	EN 15167-1 GGBFS for Use in Concrete, Mortar and Grout. Definitions, Specifications and Conformity Criteria

name of the mineral admixture used in CEM II. There are seven different
types of CEM II as portland-slag, portland-SF, portland-pozzolan,
portland-burnt shale, portland-limestone and portland-composite cements.
All the cements with mineral admixtures as main constituents are further
grouped in themselves according to the amount of mineral admixtures used
as A, B or C where, A < B < C.

Major constituents and their designations used in the standard are given
in Table 13.2. Besides these, cements may contain minor additional constituent (mac), calcium sulphates and chemical admixtures. Minor additional

Table 13.2 Major constituents of cements according
to EN 197-1

Major constituent	Designation
PC clinker	K
BFS	S
SF	D
Natural pozzolan	P
Natural calcined pozzolan	Q
Siliceous (low-lime) fly ash	V
Calcareous (high-lime) fly ash	W
Calcined shale	T
Limestone (organic carbon content <0.5%)	L
Limestone (organic carbon content <0.2%)	LL
Mixture of two or more mineral admixtures	M

constituents are mineral matter that may be used up to 5% (by mass) of the major constituents' amount. Chemical admixtures are used to facilitate cement production and/or to improve various properties of the end product. Their amount should not exceed 1% of the cement, by mass. Calcium sulphate in gypsum, hemihydrate or anhydrite forms is added as a percentage of the major constituent + mac mixture. Relative proportions of the major constituents in the 27 cements described in EN 197-1 are given in Table 13.3.

Requirements of EN 197-1 for the mineral admixtures to be used as cement constituents are listed in Table 13.4. Several terms in the table require explanation:

Reactive SiO₂ is that portion of silica which is dissolved after being treated with HCl and then in boiling KOH solution.

Reactive CaO is that portion of lime that would react to form calcium silicate hydrates and calcium aluminate hydrates.

LoI of FA mainly comes from the unburnt carbon. Allowable LoI ranges for fly ashes are given in three categories in order to help consumers to take necessary precautions with air-entrained concretes.

13.1.2 Requirements for mineral admixtures in ASTM C 595

The ASTM standard specification for blended cements which was released in 2014 pertains to hydraulic cements obtained by slag, pozzolan, limestone or combinations of these with PC or slag with lime. The standard specifies two major groups of blended cements as (a) binary blended cements which consist of PC with either a slag, a pozzolan or a limestone and (b) ternary blended cements which consist of PC with either a combination of two different pozzolans, a slag and a pozzolan, a slag and a limestone or a pozzolan and a limestone. Binary blended cements are designated as Type IS (portland-BFS cement), Type IP (portland-pozzolan cement), Type IL (portland-limestone cement) and ternary blended cements are designated by Type IT. The naming practice of blended cements is made by adding a suffix (X) which gives the percentage of slag, pozzolan or limestone in the binary blends and by adding suffixes (AX) and (BY) to ternary blend designation. A and B stand for either S (slag) or P (pozzolan) or L (limestone) and X and Y are the percentages of A and B, respectively. Requirements of ASTM C 595 (2014) for the mineral admixtures to be used as blended cement constituents are listed in Table 13.5.

13.1.3 Comparison of chemical and physical requirements for blended cements in EN 197-1 and ASTM C 595

EN 197-1 specifies three strength classes for cements as 32.5, 42.5 and 52.5. Each of these values corresponds to the minimum 28-day compressive

Table 13.3 Mass proportions of major constituents in EN 197-1 cements

Cement type			Mass proportions of major constituents (%)									
			Clinker	Slag	Silica fume	Natural pozzolan		Fly ash		Calcined shale	Limestone	
Major cement type	Name	Designation	K	S	D	P	Q	V	W	T	L	LL
CEM I	Portland cement	CEM I	95–100	–	–	–	–	–	–	–	–	–
CEM II	Portland-slag cement	CEM II/A-S	80–94	6–20	–	–	–	–	–	–	–	–
		CEM II/B-S	65–79	21–35	–	–	–	–	–	–	–	–
	Portland-silica fume cement	CEM II/A-D	90–94	–	6–10	–	–	–	–	–	–	–
	Portland-pozzolan cement	CEM II/A-P	80–94	–	–	6–20	–	–	–	–	–	–
		CEM II/B-P	65–79	–	–	21–35	–	–	–	–	–	–
		CEM II/A-Q	80–94	–	–	–	6–20	–	–	–	–	–
		CEM II/B-Q	65–79	–	–	–	21–35	–	–	–	–	–
	Portland-fly ash cement	CEM II/A-V	80–94	–	–	–	–	6–20	–	–	–	–
		CEM II/B-V	65–79	–	–	–	–	21–35	–	–	–	–
		CEM II/A-W	80–94	–	–	–	–	–	6–20	–	–	–
		CEM II/B-W	65–79	–	–	–	–	–	21–35	–	–	–
	Portland-calcined shale cement	CEM II/A-T	80–94	–	–	–	–	–	–	6–20	–	–
		CEM II/B-T	65–79	–	–	–	–	–	–	21–35	–	–

(Continued)

Table 13.3 (Continued) Mass proportions of major constituents in EN 197-1 cements

Major cement type	Cement type Name	Designation	Clinker K	Slag S	Silica fume D	Natural pozzolan P	Natural pozzolan Q	Fly ash V	Fly ash W	Calcined shale T	Limestone L	Limestone LL
	Portland-limestone cement	CEM II/A-L	80–94	–	–	–	–	–	–	–	6–20	–
		CEM II/B-L	65–79	–	–	–	–	–	–	–	21–35	–
		CEM II/A-LL	80–94	–	–	–	–	–	–	–	–	6–20
		CEM II/B-LL	65–79	–	–	–	–	–	–	–	–	21–35
	Portland-composite cement	CEM II/A-M	80–88	Any two or more, 12–20							
		CEM II/B-M	65–79	Any two or more, 21–35							
CEM III	Blast furnace slag cement	CEM III/A	35–64	36–65	–	–	–	–	–	–	–	–
		CEM III/B	20–34	66–80	–	–	–	–	–	–	–	–
		CEM III/C	5–19	81–95	–	–	–	–	–	–	–	–
CEM IV	Pozzolanic cement	CEM IV/A	65–89	–	Any two or more, 11–35						–	–
		CEM IV/B	45–64	–	Any two or more, 36–55						–	–
CEM V	Composite cement	CEM V/A	40–64	18–30	–	Any one or more, 18–30			–	–	–
		CEM V/B	20–38	31–49	–	Any one or more, 31–49			–	–	–

Table 13.4 EN 197-1 (2012) requirements for mineral admixtures used in cements

Mineral admixture		
Designation	Name	Requirements
S	GBFS	• Glassy phase content ≥2/3 (by mass) • CaO + MgO + SiO₂ ≥2/3 (by mass) • (CaO + MgO)/SiO₂ ≥1.0 (by mass)
P	Natural pozzolan	• Reactive SiO₂ ≥25% (by mass)
Q	Natural calcined pozzolan	

V	Siliceous FA	• Reactive SiO₂ ≥25% (by mass)
		• Reactive CaO <10% (by mass)
		• Free CaO <1% (by mass)
		• Free CaO = 1.0%–2.5% is acceptable if Le Chatelier expansion of 30 : 70 (by mass) FA : CEM I paste is <10 mm
		• LoI may be between 0% and 5% (category A)
		  2%–7% (category B)
		  4%–9% (category C)

W	Calcareous FA	• Reactive CaO ≥10% (by mass) • For reactive CaO = 10%–15%, Reactive SiO₂ ≥25% • For reactive CaO >15%, 28-day compressive strength of FA mortar ≥10 MPa • Le Chatelier expansion of 30 : 70 (by mass) FA : CEM I paste should be <10 mm
T	Calcined shale	• 28-day compressive strength of calcined shale mortar ≥10 MPa • Le Chatelier expansion of 30 : 70 (by mass) calcined shale : CEM I paste should be <10 mm
L, LL	Limestone	• CaCO₃ ≥75% (by mass) • Clay content ≤1.2 g/100 g • Total organic carbon content (TOC) ≤0.5% for L ≤0.2% for LL
D	SF	• Amorphous SiO₂ ≥85% (by mass) • Elemental Si ≤0.4% (by mass) • LoI ≤4% (by mass) • Specific surface area (BET) ≥15,000 m²/kg

Table 13.5 ASTM C 595 (2014) requirements for mineral admixtures used in blended cements

Mineral admixture	Requirements
Pozzolan	• Amount retained on 45 μm sieve (wet sieved) ≤20% • Alkali reactivity, mortar bar expansion at 91 days ≤0.05% (for blended cements containing less than 15% pozzolan) • 28-day activity index with PC ≥75% • LoI ≤10% (for natural pozzolan) ≤6% (for FA or SF)
Slag	• Amount retained on 45 μm sieve (wet sieved) ≤20% • 28-day activity index with PC ≥75%
Limestone	• CaCO₃ ≥70% • Methylene blue index ≤1.2 g/100 g • Total organic carbon content (TOC) ≤0.5%

strength, in MPa. Each strength class is subdivided into two groups according to the early (2-days) strength requirements as N and R, which stand
for normal and high-early strength, respectively. Furthermore, CEM III
cements may have a third strength group as L which stands for lower
early strength. ASTM C 595, on the other hand, specifies only minimum
3-, 7- and 28-day compressive strengths for two groups of blended cements
according to their slag contents. The strength requirements of both standards are summarized in Table 13.6. It should be noted that, compressive
strength values specified in EN 197-1 are determined on specimens with
constant water–cementitious ratio of 0.50 according to EN 196-1 (2005)
whereas those in ASTM C 595 are determined on specimens with water–
cementitious ratios that give $110 \pm 5\%$ flow according to ASTM C 109
(2013). If special properties like sulphate resistance, heat of hydration or
air-entrainment are required from the blended cements in ASTM C 595
then the strength requirements given in Table 13.6 change as in Table 13.7.

There are three main types of sulphate resisting cements in EN 197-1
(2012): (1) sulphate resisting PC (CEM I-SR), (2) sulphate resisting BFS
cement (CEM III-SR) and (3) sulphate resisting pozzolanic cement (CEM
IV-SR). CEM I-SR cement has three subgroups, depending on the maximum C_3A content of the clinker as 0%, 3% and 5%. Among the three
subgroups of BFS cements, CEM III/B and CEM III/C cements are considered as sulphate resisting but CEM III/A is not. Both pozzolanic cements,
CEM IV/A and CEM IV/B, are considered as sulphate resisting cements.
Although there is no restriction to the C_3A content of the clinker used in
CEM III/B and CEM III/C, the clinker of sulphate resisting CEM IV must
have maximum 9% C_3A. ASTM C 595 does not have such restrictions as
long as the maximum sulphate expansion determined by ASTM C 1012
(2013) does not exceed 0.10% at 180 days for moderate sulphate resistance
and 0.05% at 6 months and 0.10% at one year for high sulphate resistance.

EN 197-1 (2012) defines low-heat cements as those which have maximum 270 J/g heat of hydration in 7 days, determined by heat of solution
method (EN 196-8, 2011) or at 41 h, determined by semi-adiabatic method
(EN 196-9, 2010). ASTM C 595, on the other hand, has both 7-day and
28-day requirements. 290 J/g and 330 J/g for moderate heat of hydration
and 250 J/g and 290 J/g for low heat of hydration as determined by heat of
solution method described in ASTM C 186 (2015) are specified as maximum heat of hydration values at 7 and 28 days, respectively.

Chemical and physical requirements of both standards are given in
Tables 13.8 and 13.9, respectively.

13.1.4 ASTM C 1157

Cements in this standard are designated Type GU (general use), Type HE
(high-early-strength), Type MS (moderate sulphate resistance), Type HS
(high sulphate resistance), Type MH (moderate heat of hydration), Type LH

Table 13.6 Strength requirements of EN 197-1 and ASTM C 595, for blended cements

Test age (days)	EN 197-1 blended cements									ASTM C 595 blended cements	
	32.5L	32.5N	32.5R	42.5L	42.5N	42.5R	52.5L	52.5N	52.5R	IL IP IS(<70) IT(S <70)	IS(≥70) IT(S ≥70)
2	–	–	≥10	–	≥10	≥20	≥10	≥20	≥30	–	–
3	–	–	–	–	–	–	–	–	–	≥13	–
7	≥12	≥16	–	≥16	–	–	–	–	–	≥20	≥5
28		≥32.5, ≤52.5			≥42.5, ≤62.5			≥52.5		≥25	≥11

Table 13.7 Strength requirements of ASTM C 595 for blended cements with special requirements

Test age (days)	Blended cements with special properties				
	Air-entrained	Moderate sulphate resistance	High sulphate resistance	Moderate heat of hydration	Low heat of hydration
3	≥10.4	≥11.0	≥11.0	≥10.4	–
7	≥16.0	≥18.0	≥18.0	≥16.0	≥11.0
28	≥20.0	≥25.0	≥25.0	≥20.0	≥21.0

(low heat of hydration). All six types may have Option A (air-entrained) or Option R (low reactivity with alkali-reactive aggregates). Type GU cements are used when the special properties of the other five types are not required. The performance requirements of cements specified in ASTM C 1157 may be attained with or without mineral admixture incorporation.

ASTM C 1157 has the same requirements for soundness, setting time and 14-day mortar bar expansion for all cements. Maximum autoclave expansion is 0.80%, minimum and maximum allowable initial setting times are 45 min. and 420 min., respectively, and maximum mortar bar expansion at 14 days is 0.02%. Strength requirements differ for each type. Those of types MS and HS are slightly lower when compared with type GU. Types MH and LH have even lower strength requirements. Type HE is the only cement in the standard that has a 1-day strength requirement.

13.2 OVERVIEW OF STANDARDS ON MINERAL ADMIXTURES IN CONCRETE

The properties of mineral admixtures to be used as separate ingredients in concrete are specified in three European and three ASTM standards. A list of the standards is given in Table 13.10.

Chemical and physical requirements of the standards for FA, GBFS and SF are summarized in Tables 13.11 and 13.12, respectively. There are several points that should be indicated before comparing the standards' requirements: EN 450-1 specifies three LoI categories A, B and C and two fineness categories as N and S for fly ashes. While EN 450-1 considers only low-lime fly ashes suitable as concrete mineral admixtures, ASTM C 618 allows the use of high-lime fly ashes, as long as they meet the standard's requirements. Strength activity indices are determined on specimens with different mass proportions in EN 450-1 and ASTM C 618. In EN 450-1, 25% (by mass) PC is replaced by FA whereas 20% (by mass) PC is replaced by FA in ASTM C 618. GGBFS : PC proportions are 50 : 50 (by mass) for strength activity specimens in both EN 15167-1 and ASTM C 989. For the

Table 13.8 Chemical requirements for blended cements in ASTM C 595 (2014) and EN 197-1 (2012)

	ASTM C 595 blended cements				EN 197-1 blended cements						
					CEM II		CEM III	CEM IV		CEM V	
	IS(<70) IT(P < S < 70) IT(L < S < 70)	IS(≥70) IT(S ≥70)	IP IT(P ≥S) IT(P ≥L)	IL IT(L ≥S) IT(L ≥P)	32.5N 32.5R 42.5N	42.5R 52.5N 52.5R	All	32.5N 32.5R 42.5N	42.5R 52.5N 52.5R	32.5N 32.5R 42.5N	42.5R 52.5N 52.5R
MgO, %	–	–	≤6.0	–	–	–	–	–	–	–	–
SO_3, %	≤3.0	≤4.0	≤4.0	≤3.0	≤3.5	≤4.0	≤4.0	≤3.5	≤4.0	≤3.5	≤4.0
S^{2-}, %	≤2.0	≤2.0	–	–	–	–	–	–	–	–	–
Ins. Res., %	≤1.0	≤1.0	–	–	–	–	≤5.0	–	–	–	–
LoI, %	≤3.0	≤4.0	≤5.0	≤10.0	–	–	≤5.0	–	–	–	–
Cl^-, %	–	–	–	–	≤0.1	≤0.1	≤0.1	≤0.1	≤0.1	≤0.1	≤0.1

Table 13.9 Physical requirements for blended cements in ASTM C 595 (2014) and EN 197-1 (2012)

	ASTM C 595 blended cements	EN 197-1 blended cements											
		CEM II			CEM III			CEM IV			CEM V		
		32.5	42.5	52.5	32.5	42.5	52.5	32.5	42.5	52.5	32.5	42.5	52.5
Autoclave expansion (%)	≤0.80		—			—			—			—	
Autoclave cont., %	≤0.20		—			—			—			—	
Le Chatelier expansion (mm)	—		≤10			≤10			≤10			≤10	
Initial setting time (min)	≥45	≥75	≥60	≥45	≥75	≥60	≥45	≥75	≥60	≥45	≥75	≥60	≥45
Initial setting time (h)	≤7		—			—			—			—	
Air content of mortar (vol. %)	12		—			—			—			—	

Table 13.10 EN and ASTM standards on mineral admixtures for concrete

Mineral admixture	European standard	ASTM standard
Natural pozzolan, raw or calcined	–	ASTM C 618
Fly ash	EN 450-1	ASTM C 618
Ground granulated blast furnace slag	EN 15167-1	ASTM C 989
Silica fume	EN 13263-1	ASTM C 1240

case of SF, the proportions are 10 : 90 (by mass) in both EN 13263-1 and ASTM C 1240.

ASTM C 989 specifies three grades of slag as Grade 80, Grade 100 and Grade 120 according to their performance in strength activity test.

EN 13263-1 specifies two categories of SF as S1 and S2, according to their total SiO_2 contents.

Table 13.11 Chemical requirements of relevant EN and ASTM standards for mineral admixtures in concrete

Requirement	Natural pozzolan		Fly ash (ASTM C 618, 2012)		Ground granulated blast furnace slag		Silica fume	
	(ASTM C 618, 2012)	(EN 450-1, 2015)	Type F	Type C	(EN 15167-1, 2006)	(ASTM C 989, 2014)	(EN 13263-1, 2010)	(ASTM C 1240, 2014)
$SiO_2 + Al_2O_3 + Fe_2O_3$ (%)	≥70.0	≥70.0	≥70.0	≥50.0				
SO_3 (%)	≤4.0	≤3.0	≤5.0	≤5.0	≤2.5	≤4.0	≤2.0	
Moisture content (%)	≤3.0		≤3.0	≤3.0				
LoI (%)	≤10.0	A: ≤5.0 B: 2.0–7.0 C: 4.0–9.0	≤6.0	≤6.0	≤3.0		≤4.0	≤6.0
Cl^- (%)		≤0.1			≤0.1		≤0.3	
S (%)					≤2.0	≤2.5		
Free CaO (%)		≤1.0					≤1.0	
Reactive CaO (%)		≤10.0						
Reactive SiO_2 (%)		≥25.0						
Total SiO_2 (%)							S1: ≥85.0 S2: ≥80.0	≥85.0
Elemental Si (%)							≤0.4	
CaO + MgO + SiO_2 (Mass fraction)					≥2/3			
(CaO + MgO)/SiO_2					≥1.0			
Total alkali (%) $(Na_2O)_{eq}$		≤5.0						
MgO (%)		≤4.0				≤18.0		
Soluble P_2O_5 (mg/kg)		≤100.0						

Table 13.12 Physical requirements of relevant EN and ASTM standards for mineral admixtures in concrete

Requirement	Natural pozzolan		Fly ash (ASTM C 618, 2012)		Ground granulated blast furnace slag		Silica fume	
	(ASTM C 618, 2012)	(EN 450-1, 2015)	Type F	Type C	(EN 15167-1, 2006)	(ASTM C 989, 2014)	(EN 13263-1, 2010)	(ASTM C 1240, 2014)
% Retained on 45 μm sieve (wet sieved)	≤34.0	N: ≤40.0 S: ≤12.0	≤34.0	≤34.0		≤20.0	≤10.0	
Blaine fineness (m²/kg)					≥275			
BET fineness (m²/g)							15.0–35.0	≥15.0
Strength acitivity index with portland cement (% of control)	7-day: ≥75.0 28-day: ≥75.0	28-day: ≥75.0 90-day: ≥85.0	7-day: ≥75.0 28-day: ≥75.0	7-day: ≥75.0 28-day: ≥75.0	7-day: ≥45.0 28-day: ≥70.0	**Grade 80:** – **Grade 100:** 7-day: ≥75.0 28-day: ≥75.0 **Grade 120:** 7-day: ≥95.0 28-day: ≥115.0	28-day: ≥100.0	7-day: ≥105.0
Le Chatelier expansion (mm)					≤10.0			
Autoclave expansion (%)	≤0.8		≤0.8	≤0.8				
Water requirement (% of control)	≤115.0	N: – S: ≤95.0	≤105.0	≤105.0				

Chapter 14

Use of mineral admixtures in special concretes

Although it is a versatile material, conventional concrete may have certain limitations for various special applications and conditions and may not be able to meet the requirements of some stringent specifications. Studies for certain modifications and improvements in conventional concrete have led to various special types of concretes. As in conventional concretes, it is possible or sometimes necessary to use mineral admixtures in many special types of concretes. Several special concretes are briefly described in this chapter with emphasis on their significant properties and the effects of the use of mineral admixtures on those properties.

14.1 LIGHTWEIGHT CONCRETE

Lightweight concretes can be divided into two classes as structural and non-structural. Structural lightweight concretes are the ones with dry unit weight less than 1,800 kg/m³ and 28-day compressive strength of at least 17 MPa. Non-structural lightweight concretes may further be divided into two as moderate strength lightweight concretes and low-density lightweight concretes. The former usually have a 28-day compressive strength between 7 and 15 MPa whereas those of the latter are usually less than 3 MPa. Dry unit weights of non-structural lightweight concretes are less than 1,400 kg/m³ and less than 800 kg/m³, respectively (ACI 213R, 2014). Lightweight concretes can be obtained (a) using lightweight aggregates (lightweight aggregate concrete), (b) by introducing gas bubbles into concrete (gas concrete, aerated concrete, foamed concrete or cellular concrete) or (c) using only coarse aggregates (no-fines concrete).

Many different natural, processed natural or synthetic aggregates may be used to obtain lightweight concrete. The common feature of such aggregates is low density due to their high internal porosity. Dry unit weights of various lightweight aggregates are given in Figure 14.1.

The strength of lightweight aggregate concretes depends largely on the type and volume of aggregate used. Very porous aggregates located at the left-hand side of Figure 14.1 are weak and therefore they are used mostly

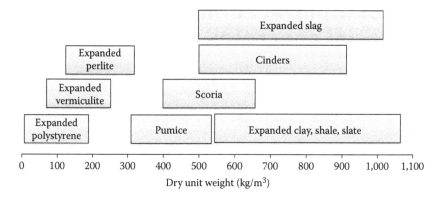

Figure 14.1 Dry unit weight ranges for various lightweight aggregates.

in making insulating concrete. Low water–cement ratios are necessary for achieving sufficient strength in structural lightweight concretes. However, high absorption (10%–20%) of lightweight aggregates makes it very difficult to calculate the exact mixing water content.

When compared with conventional concrete, the workability of lightweight aggregate concrete requires more attention due to several reasons: (1) high porosity and generally rough surface texture of lightweight aggregates necessitate more paste; (2) some lightweight aggregates may continue to absorb water after mixing which may lead to serious slump losses and (3) high slump and prolonged vibration may easily lead to floating of lightweight coarse aggregates.

Lightweight aggregate concretes have lower moduli of elasticity and higher cement paste volume than conventional concretes. Therefore, they are generally more susceptible to creep and shrinkage deformations.

Durability issues in lightweight aggregate concretes are similar to those in normal concretes. High absorption of the aggregates does not mean high permeability of the concrete. Usually, the measures taken to improve the durability of conventional concretes also hold true for lightweight aggregate concretes. On the other hand, since most lightweight aggregates are more easily breakable than normal weight aggregates, lightweight concretes are not as resistant as normal concretes to wear and abrasion.

Although there is not much research work published on the use of mineral admixtures in lightweight aggregate concrete, several are addressed here.

Effects of FA, GGBFS and SF on the workability and strength of high-slump and high-strength lightweight aggregate concretes were investigated by Chen and Liu (2008). Normal weight sand and expanded clay were used as aggregates. They used 10, 20, 30 and 40% (by mass) FA and blast furnace slag and 5, 10 and 15% (by mass) SF to partially replace PC. Additionally, ternary mixtures of 5% SF and 10% FA or blast furnace slag, 15% FA and 15% blast furnace slag and a quaternary mixture consisting of 10% of each mineral admixture were prepared. Total cementitious material, water–cementitious ratio, superplasticiser content, sand and expanded clay contents were kept constant as 530 kg/m³, 0.30, 1.5% (by mass of cementitious material), 265 kg/m³ and 400 kg/m³, respectively. Slumps of all mixes were measured right after mixing and 30 and 60 min later. Initial slump and slump flow increased with increasing FA content, decreased with increasing SF content and did not show significant changes with blast furnace slag, as shown in Figure 14.2. Although slump increased with increasing FA content, bleeding and up-floating aggregates resulted in non-uniformity for 30% and 40% FA incorporation. The control concrete without any mineral admixture had slump losses of 3% and 25% after 30 and 60 min, respectively. Slump losses of blast furnace slag-incorporated concretes were similar to those of the control (3.1% to 7.6% after 30 min and 20% to 25% after 60 min). On the other hand, slump losses of FA concretes ranged between 7% and 18% after 30 min and 30% and 36% after 60 min. SF-incorporated concretes experienced the highest slump losses (33%–47% after 30 min and 50%–56% after 60 min).

The control lightweight aggregate concrete had a 50 MPa 28-day compressive strength. FA incorporation resulted in lower 7-day strengths than the control. The reduction was not significant for 10% and 20% FA. 28-day strengths of these two concretes were about 5% and 14% higher than that of the control. 30% and 40% FA concretes had slightly lower 28-day strengths. Blast furnace slag- and SF-incorporated lightweight aggregate concretes and the quaternary blends all had higher strengths than the control at both ages. Chen and Liu (2008) recommended the use of quaternary blends of PC, FA, blast furnace slag and SF as the binding medium of lightweight aggregate concretes from workability and strength points of view, as well as economy.

Kılıç et al. (2003) determined the compressive and flexural strengths of four lightweight aggregate concretes made by scoria. Three of the concretes prepared contained 10% SF, 20% high-lime fly ash and 10% SF + 20% high-lime fly ash to replace an equal mass of PC. Total cementitious material contents and water–cementitious ratios of the mixes were 500 kg/m³ and 0.55, respectively. Same aggregate gradation and amount were used in all mixes. Slump of the control mix was 70 ± 20 mm whereas those of SF, FA and SF + FA incorporated mixes were 50 ± 15, 60 ± 15 and 60 ± 25 mm, respectively. All concretes had air-dry density of 1,800–1,860 kg/m³. The density decreased with increasing amount of mineral admixture. Compressive strengths of SF concretes were higher than those of the control concrete at all ages whereas high-lime fly ash resulted in slight strength reductions at early ages but exceeded those of the control beyond 28 days. Strength of the ternary mix was in between those of the binary ones and higher than the control, at all ages.

In a more recent study, fine and coarse scoria and normal weight sand were used as aggregate in concrete mixes having a constant cementitious materials content of 400 kg/m³ to obtain structural lightweight concretes with 28-day strength of 30 MPa. SF and FA were used to partially replace the PC. Ten binary and ternary mixes were prepared besides the control.

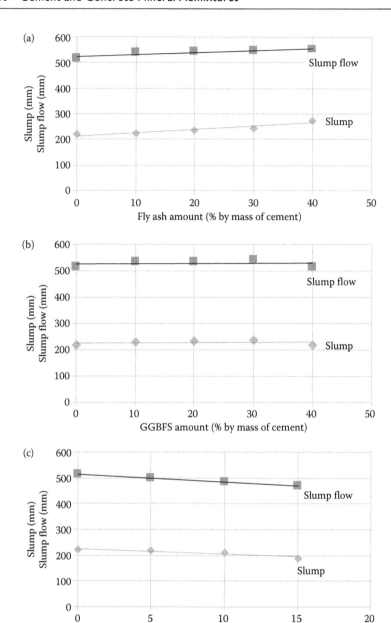

Figure 14.2 Changes in initial slump and slump flow in mineral admixture-incorporated lightweight aggregate concretes: (a) Fly ash, (b) GGBFS and (c) SF. (Data from Chen, B., Liu, J. 2008. *Construction and Building Materials*, 22, 655–659.)

Table 14.1 Mix proportions of structural lightweight aggregate concretes used by Shannag (2011)

Mix no.	Cement (kg/m³)	Fly ash (kg/m³)	Silica fume (kg/m³)	Natural sand (kg/m³)	Coarse scoria (kg/m³)	Fine scoria (kg/m³)	Water (kg/m³)	Superplasticiser (l/m³)
C	400	–	–					1
SF1	380	–	20					3
SF2	360	–	40					3
SF3	340	–	60					4
FA1	380	20	–					2
FA2	360	40	–	198–200	543–550	345–350	247–250	1.5
FASF1	360	20	20					3
FASF2	340	40	20					2
SFFA1	340	20	40					4
SFFA2	320	40	40					4
SFFA3	320	20	60					3

All mixes had a constant water–cementitious ratio of 0.625. In order to attain similar workability, superplasticiser amounts were adjusted (Shannag, 2011). Slump values ranged from 90 to 180 mm. Air dry densities were between 1,935 and 1,995 kg/m³. Concrete mix proportions are given in Table 14.1 and 28-day compressive strength and modulus of elasticity values are shown in Figure 14.3.

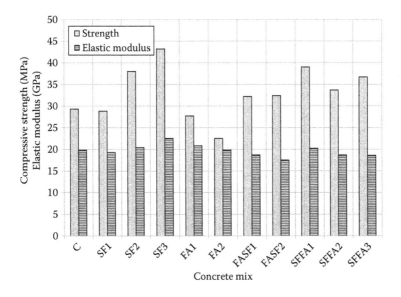

Figure 14.3 28-day compressive strength and elastic modulus values of structural lightweight aggregate concretes prepared by Shannag (2011).

Studies on much lighter weight concretes such as expanded polystyrene aggregate concrete (Babu et al., 2005), compressed lightweight building blocks (Demirdağ et al., 2008) and cellular concrete (Jitchaiyaphum et al., 2011) showed that large amounts of FA can also be successfully used in such concretes.

14.2 HIGH-STRENGTH CONCRETE

A concrete is defined as high strength if its 28-day compressive strength is above 55 MPa. However, this definition may change with time and geography. In 1970s, concretes with compressive strengths exceeding 40 MPa were considered as high strength. This value, from a commercial availability point of view, may still be taken as high strength in some parts of the world and as medium strength in the other parts.

Production of concrete with high strength and appropriate workability requires a more careful materials selection than normal concrete. Unless high-early strength is required, any portland or blended cement that conforms to the standards' requirements may be used. High-strength concretes usually contain more cement than the ordinary concretes which may result in higher temperature rise. Water-reducing or high-range water-reducing chemical admixtures are almost always used in the production of high-strength concretes. Depending on other desired properties, retarding, accelerating, air-entraining admixtures may also be used. Their compatibility with the cement and other ingredients of the concrete should be carefully checked. The influence of aggregate on the strength of normal-strength concrete is basically related with the changes it makes in the water requirement. Denser matrix and interfacial zone due to low water–cement ratio of high-strength concrete impose aggregates important roles on the strength. Therefore, strong and hard aggregates with high modulus of elasticity and low thermal expansion coefficient should be preferred in producing high-strength concrete (Aïtcin, 1995; Mehta and Monteiro, 2006). Maximum size of the coarse aggregate should be reduced in high-strength concrete in order to improve the interfacial bond. Strengths about 70 MPa were attained by 20–25 mm maximum aggregate size. Strengths above 125 MPa were obtained with maximum aggregate sizes of 10–14 mm (Mehta and Monteiro, 2006). Generally, coarser fine aggregates should be preferred since high-strength concrete already contains considerable amounts of fine cementitious materials.

The use of mineral admixtures in high-strength concrete may contribute to (1) controlling of slump loss and stickiness of the mix, (2) reduction in heat evolution during hydration, (3) modifying the microstructure and (4) reducing the cost.

In a study to produce high-strength concrete with high FA contents, 25% and 45% (by mass) PC was replaced by a low-lime fly ash in concretes with water–cementitious ratios of 0.24 and 0.19. Amount of superplasticiser was adjusted to give slump values of 200–230 mm in all concretes prepared (Poon et al., 2000). Although early strengths in 3 and 7 days were reduced in FA concretes, 28-day and 90-day strengths were higher than the control for 25% cement replacement by FA and it was possible to obtain 28-day strength of about 90 MPa with 45% cement replacement. The strength development of the high-strength concretes are illustrated in Figure 14.4.

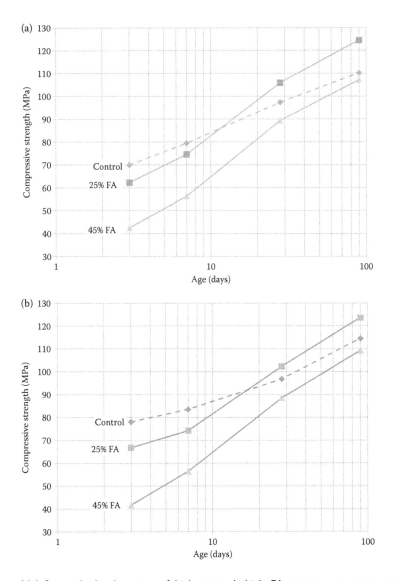

Figure 14.4 Strength development of high-strength high FA content concretes with water–cementitious ratios of: (a) 0.24 and (b) 0.19. (Data from Poon, C.S., Lam, L., Wong, Y.L. 2000. *Cement and Concrete Research*, 30, 447–455.)

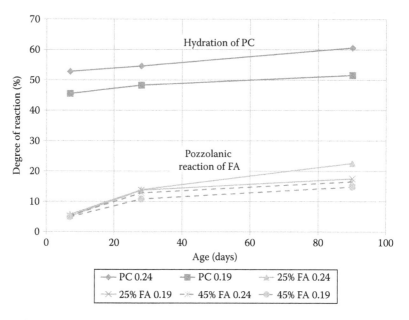

Figure 14.5 Degree of hydration and pozzolanic reaction in PC and PC-FA pastes. (Data from Poon, C.S., Lam, L., Wong, Y.L. 2000. *Cement and Concrete Research*, 30, 447–455.)

It was also found that pozzolanic reaction of the FA contributes to strength starting as early as 7 days and increases with increasing age, as illustrated in Figure 14.5. Although at 90 days of age about 40%–50% of PC and about 80%–85% of FA still remain unreacted, they would serve as micro aggregates that would also contribute to strength. Besides these, early heat evolution was considerably less in FA pastes. Within the first 3 days, total heat evolved was reduced by 15.5% and 36.3%, with respect to the control, in 25% and 45% FA specimens, respectively.

Erdem and Kırca (2008) prepared concretes of the same slump value (60 ± 10 mm) with binary blends of PC and SF and ternary blends consisting of PC + SF and low-lime fly ash, high-lime fly ash and GGBFS. Total aggregate content and grading were kept same in all mixes. Maximum aggregate size used was 25 mm. Different binder contents, ranging from 550 to 700 kg/m³, were used in five series of specimens. Superplasticiser content was 3% (by mass) of the binder and water content was adjusted to get the same slump value for all mixes. In the first stage of the experimental program binary blends consisting of 5%, 10% and 15% SF replacing the PC were investigated. Water–cementitious ratio for constant slump was determined to decrease as the total binder content increases. Increased amount of SF resulted in further slight decrease in water–cementitious ratio. Both reductions may be attributed to the increased paste volume.

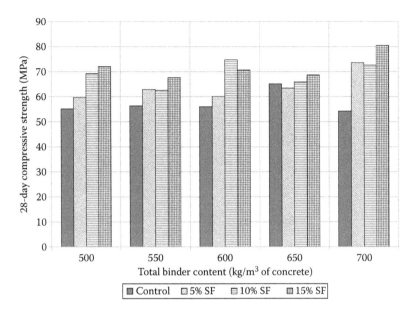

Figure 14.6 28-day compressive strengths of control and SF concretes with different total binder contents. (Data from Erdem, T.K., Kırca, Ö. 2008. *Construction and Building Materials*, 22, 1477–1483.)

SF incorporation was found to be resulting in denser and more homogeneous paste and interfacial transition zone due to (1) pozzolanic reactions that consume some of the calcium hydroxide produced upon PC hydration, (2) filler effect of SF particles that occupy the spaces between cement particles and (3) reduction in pore sizes. Increase in strength of SF concretes with respect to the control mixes are shown in Figure 14.6.

In the second part of their experimental work, Erdem and Kırca (2008) prepared ternary blends containing 10%, 20%, 30% and 40% (by mass of the total binder content) low-lime fly ash, high-lime fly ash and GGBFS with SF contents that resulted in the highest compressive strength in the first part. Compressive strength of concretes prepared using ternary blends in comparison with their SF counterparts are given in Figure 14.7. The study showed that it is possible to obtain high-strength concretes with considerable amounts of different mineral admixtures.

A similar study was carried out by Shannag (2000) on mortars made from natural pozzolan and SF blended cements and ternary blends of various SF and natural pozzolan combinations. He obtained 28-day compressive strengths between 69 and 110 MPa. The highest strength which was 26% and 60% higher than those of 15% SF and 15% natural pozzolan concretes, respectively, was attained by using a ternary blend consisting of 15% natural pozzolan and 15% SF, by mass.

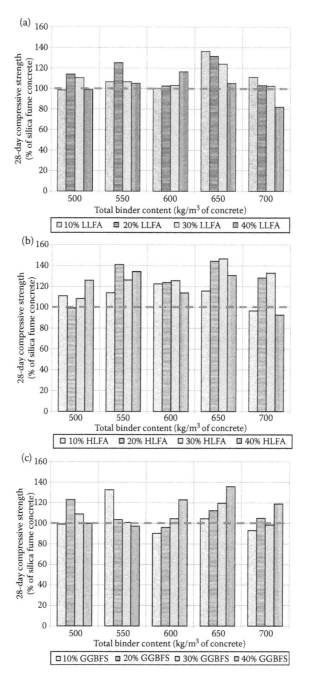

Figure 14.7 Changes in 28-day compressive strengths of concretes made by (a) low-lime fly ash, (b) high-lime fly ash and (c) GGBFS ternary blends with SF. (Data from Erdem, T.K., Kırcaf, Ö. 2008. *Construction and Building Materials*, 22, 1477–1483.)

Numerous other studies had revealed that many other mineral admixtures such as rice husk-bark ash, fluidised bed combustion fly ash, ground palm fuel oil ash (Sata et al., 2007), calcined kaolin (Poon et al., 2006; Shafiq et al., 2015), RHA (Tuan et al., 2011), and limestone powder (Isaia, et al., 2003) may be used to obtain high-strength concretes.

14.3 CONTROLLED LOW-STRENGTH MATERIALS

Controlled low-strength material (CLSM) is defined as a cementitious material having compressive strength not more than 8.3 MPa (ACI 116, 2000). There are other terms such as flowable fill, unshrinkable fill, lean-mix backfill, controlled-density fill and flowable mortar that are being used to describe CLSM. ACI Committee 229 states that it should not be considered as a type of concrete, but rather a self-compacting fill (ACI 229, 2005). CLSM may be used as backfills, structural fills, insulating and isolation fills, pavement bases, conduit bedding and for void filling purposes.

Mix proportioning of CLSM is usually done by trial and error until suitable flowability, strength and density requirements are met. Generally, cement content is 30 to 120 kg/m³. Low-lime fly ash may be as high as 1,200 kg/m³, most of which would act as an aggregate filler. High-lime fly ash may be used up to 200 kg/m³. Coarse aggregate is seldom used in CLSM. Fine aggregate content is between 1,500 and 1,800 kg/m³. Water contents typically range from 200 to 350 kg/m³. High amounts of AEA may be used in CLSMs in order to lower the density and improve the flowability. CLSMs are not normally designed to have freeze-thaw resistance. Sometimes accelerating agents may be used to accelerate hardening (ACI 229, 2005).

FA is a commonly used material in CLSM production. Other industrial by-products such as BFS (Sheen et al., 2013), bottom ash (Yan et al., 2014), spent foundry sand (Bhat and Lovell, 1997; Tikalsky et al., 2000; Siddique and Noumowe, 2008), acid mine drainage slag (Gabr and Bowders, 2000), copper slag (Taha et al., 2007) and cement kiln dust (Taha et al., 2007; Lachemi et al., 2010) has been shown to be safely used to produce CLSMs with suitable properties without any hazardous effect. Examples of CLSMs produced using only BFS or high-lime fly ash (Achtemichuk, et al., 2009) or cement kiln dust (Lachemi et al., 2008) without any PC are also available in the literature.

14.4 SELF-CONSOLIDATING CONCRETE

Self-consolidating concrete (SCC) is a concrete which has little resistance to flow so that it can easily be placed and consolidated under its own weight, yet possesses sufficient viscosity to be handled without significant segregation and bleeding. It is the plastic properties of SCC in fresh state that differentiates it from the conventional concrete (ACI 237, 2007). Generally, three workability requirements are defined for SCC: (1) *filling ability* which is the ability to flow under its own weight into all spaces within the form-work, (2) *passing ability* which is the ability to flow under its own weight through congested areas like heavily reinforced parts and (3) *stability* which is the ability to remain homogeneous during and after placing (EFNARC, 2002; ACI 237, 2007). These requirements can be obtained by balancing the deformability, stability and risk of blockage of fresh SCC. For good deformability, besides using superplasticisers, paste volume must be high, powder material must be continuously graded and coarse aggregate con-tent must be minimised. Good stability requires low water-powder ratios, using viscosity-modifying agents, reduced maximum aggregate sizes and high-specific surface area powder materials. The term powder refers to sPC and mineral admixtures. Risk of blockage can be eliminated by low coarse aggregate contents and small maximum aggregate sizes. Thus, a low yield stress from flowability and a moderate viscosity from stability points of view may be attained (Khayat, 1999).

SCC can be described as a Bingham body, the behaviour of which is mathematically given by

$$\tau - \tau_y = \eta \frac{d\gamma}{dt} \qquad (14.1)$$

where τ is the shear stress, τ_y is the yield stress, η is the plastic viscosity, and $d\gamma/dt$ is the shear rate. These rheological properties may be measured by concrete rheometers. Although a consensus on the properties determined by commercially available rheometers does not exist (Banfill, et al., 2001), they may serve for comparing the relative yield stresses and plastic viscosi-ties of different mixes. On the other hand, there are several test methods developed to measure different workability requirements for SCC some of which are listed in Table 14.2. No single test method is capable of deter-mining all the workability properties; therefore, combination of several tests becomes necessary to characterise SCC mixes.

Mineral admixtures are used either as separate ingredients or as part of blended cements in producing SCC. Limestone powder, natural pozzolans, SF, GGBFS and FA can enhance the filling and passing abilities and stabil-ity of SCC. Proper selection and proportioning of these materials would improve the packing density and reduce the water and high-range water-reducing agent requirements (ACI 237, 2007).

An investigation that considered the effect of a low-lime fly ash and a GGBFS on flowability and passing ability of SCC is summarised below. FA and GGBFS were used to replace 20, 40 and 60% (by mass) PC in preparing SCCs. Total cementitious material content and high-range water-reducing agent dosage were kept constant. Water–cementitious ratio ranged between 0.34 and 0.30. Three series of SCCs containing 0, 5 and 10% (by mass) very fine calcite powder partially replacing the aggregate were produced by

Table 14.2 Test methods for measuring SCC characteristics

Characteristic	Test method	Description	Common ranges for SCC
Flowability/ Filling ability	Slump flow	Test is similar to conventional slump test but the average spread of SCC is measured horizontally instead of the vertical slump value.	550–850 mm
Flowability/ Viscosity	T_{500} slump flow	Time required for SCC to spread 500 mm is measured.	2–5 s
	V-funnel flow time	A V-shaped funnel is filled with fresh SCC and the time taken for it to flow out of the funnel is measured.	6–20 s
Passing ability	L-box	SCC is filled into the vertical part of an L-shaped box with 2 or 3 steel bars vertically placed between the vertical and horizontal sections and the gate between the two sections is opened. Depths of SCC at the end of the horizontal section (H_2) and right behind the gate (H_1) are measured when movement of SCC stops. Passing ability or blocking ratio is calculated as $PA = H_2/H_1$	0.8–1.0
	J-ring	A ring of reinforcement bars is placed around the slump cone. The cone filled with SCC is removed, final spread is measured and the difference between the slump flow and J-ring slump flow is calculated.	0–50 mm
Stability	Sieve segregation	SCC is poured onto a 5 mm sieve. It is allowed to stand for 2 min. The mass that passed through the sieve is measured and the segregation ratio (SR) is calculated as the ratio of SCC passing through the sieve to the original mass.	≤15%
	Column segregation	SCC is poured into a cylinder of 600 mm height and let stand for 15 min. Then several sections are removed and washed over No. 4 or 5 mm sieve and the retained aggregate is weighed.	≤10%

Source: Adapted from ACI 237. 2007. Self-consolidating concrete, ACI Committee 237 Report, ACI 237R-07, American Concrete Institute, Farmington Hills, MI; EFNARC, 2002.

using each of the mineral admixtures. Calcite powder was used as a viscosity-modifying agent. Maximum aggregate size was chosen as 12.5 mm. Slump flow, T_{500}, V-funnel, J-ring and L-box tests were carried out on fresh SCCs and strength tests at 7, 28 and 90 days were carried out on hardened SCCs (Beycioğlu and Aruntaş, 2014). Workability properties are given in Figures 14.8–14.12, respectively.

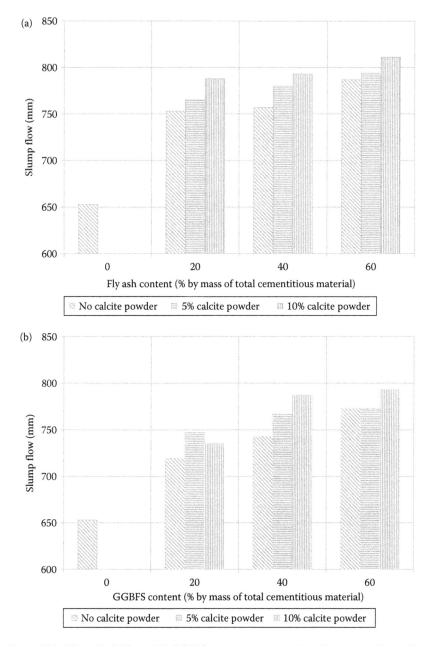

Figure 14.8 Effect of (a) FA and (b) GGBFS content on slump flow. (Data from Beycioğlu, A., Aruntaş, H.Y. 2014. *Construction and Building Materials*, 73, 626–635.)

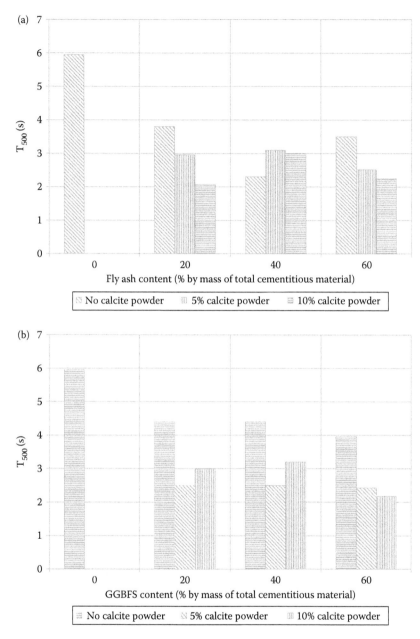

Figure 14.9 Effect of (a) FA and (b) GGBFS content on T_{500}. (Data from Beycioğlu, A., Aruntaş, H.Y. 2014. *Construction and Building Materials*, 73, 626–635.)

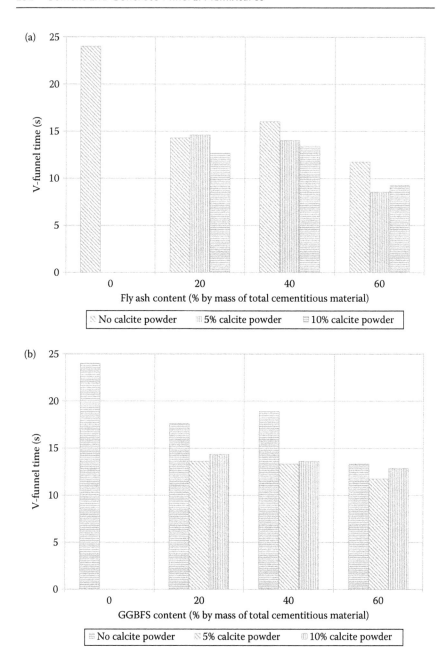

Figure 14.10 Effect of (a) FA and (b) GGBFS content on V-funnel time. (Data from Beycioğlu, A., Aruntaş, H.Y. 2014. *Construction and Building Materials*, 73, 626–635.)

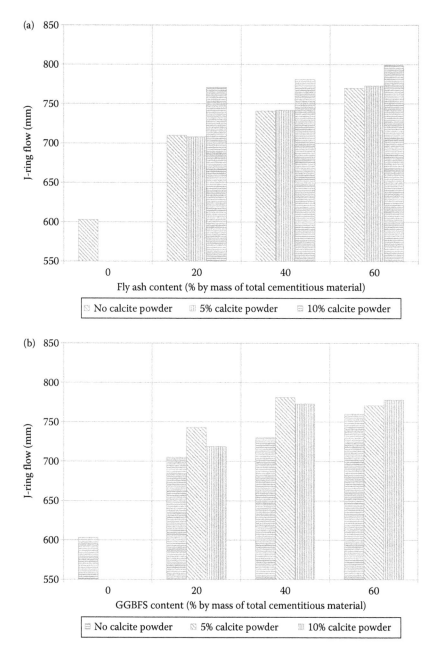

Figure 14.11 Effect of (a) FA and (b) GGBFS content on J-ring flow. (Data from Beycioğlu, A., Aruntaş, H.Y. 2014. *Construction and Building Materials*, 73, 626–635.)

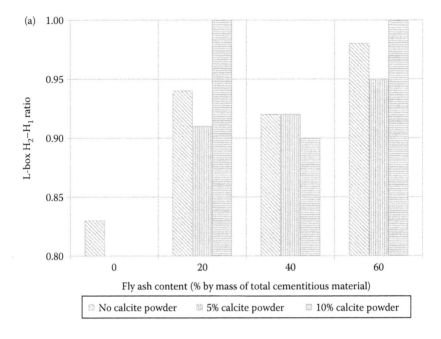

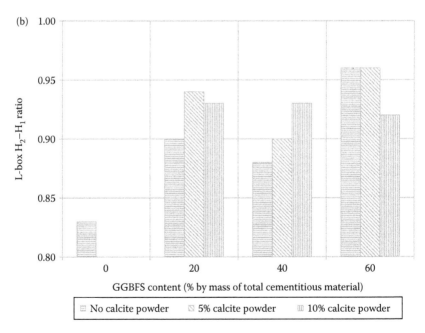

Figure 14.12 Effect of (a) FA and (b) GGBFS content on L-box H_2–H_1 ratio. (Data from Beycioğlu, A., Aruntaş, H.Y. 2014. *Construction and Building Materials*, 73, 626–635.)

Although Beycioğlu and Aruntaş (2014) were not able to give expressions relating the mineral admixture content to the properties investigated, their results clearly indicate that both mineral admixtures they used improved the flowability and passing ability of SCC and FA performed better than GGBFS. On the other hand, FA and GGBFS resulted in lower strengths than the control.

Şahmaran et al. (2006) investigated the effects of FA, limestone powder, brick powder and MK on filling ability and viscosity of self-consolidating mortars by mini slump and mini V-funnel tests and determined that FA and limestone powder improve these properties, whereas brick powder and MK are not as effective. Setting times were increased and strengths were reduced upon mineral admixture incorporation.

In another study in which MK was used to partially replace either PC or limestone powder upto 20% and 40%, respectively, slump flow and V-funnel time increased and L-box height ratio decreased with respect to those of the control indicating that the viscosity of the mixes become higher (Sfikas et al., 2014).

Benaicha et al. (2015) replaced 5%–30% (by mass) PC by SF in SCC mixtures. The control mix was prepared using 350 kg/m^3 PC and 170 kg/m^3 limestone powder. Water–cementitious ratio, superplasticiser and fine and coarse aggregate contents were kept constant in all mixes as 0.37, 7.8 kg/m^3, 890 kg/m^3 and 900 kg/m^3, respectively. Slump flow, L-box H$_2$–H$_1$ ratio and sieve segregation decreased and V-funnel flow time increased with increasing SF content as shown in Figures 14.13–14.16. In other words, flowability and passing ability decreases and viscosity increases with increasing amount of SF. Indeed, rheometer test results revealed that both the yield stress and plastic viscosity increased with increasing amount of SF in the mix, as shown in Figure 14.17. In a similar study by Lu et al. (2015) SCCs prepared with PC and 2%–16% (by mass) SF were tested by slump flow, J-ring flow and rheometer tests. Flowability and passing ability of SF-incorporated SCCs were found to decrease with increasing SF content. On the other hand, rheometer test results in this investigation showed that yield stress increased with increase in SF content whereas plastic viscosity decreased upto 4% SF and increased afterwards but still had values less than that of the control, upto 12% SF.

14.5 FIBRE-REINFORCED CONCRETE

Concrete is a brittle material with low tensile strength. Most of the time conventional steel reinforcement is relied from ductility and tensile load carrying points of view. Cracking caused by tensile stresses developed in concrete exceeding the tensile strength would lead to decreased mechanical and durability properties. So, any improvement in the ductility and tensile strength of concrete is valuable. One such technique is based on a method

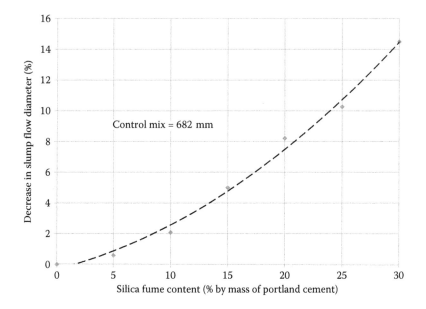

Figure 14.13 Effect of SF content on slump flow of SCCs. (Data from Benaicha, M. et al. 2015. *Construction and Building Materials*, 84, 103–110.)

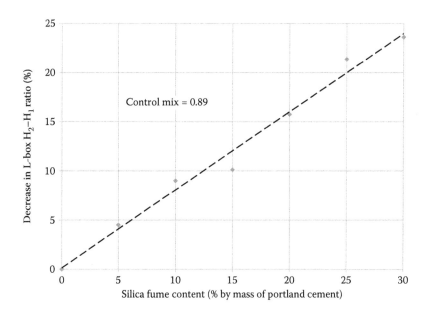

Figure 14.14 Effect of SF content on L-box H_2–H_1 ratio of SCCs. (Data from Benaicha, M. et al. 2015. *Construction and Building Materials*, 84, 103–110.)

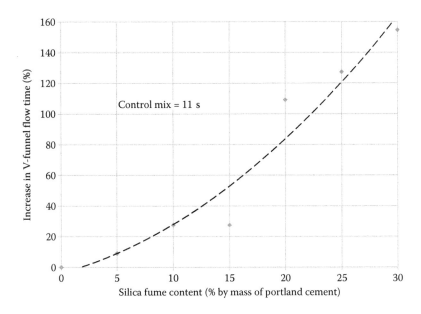

Figure 14.15 Effect of SF content on V-funnel flow time of SCCs. (Data from Benaicha, M. et al. 2015. *Construction and Building Materials*, 84, 103–110.)

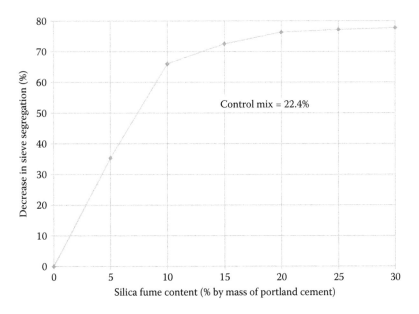

Figure 14.16 Effect of SF content on sieve segregation of SCCs. (Data from Benaicha, M. et al. 2015. *Construction and Building Materials*, 84, 103–110.)

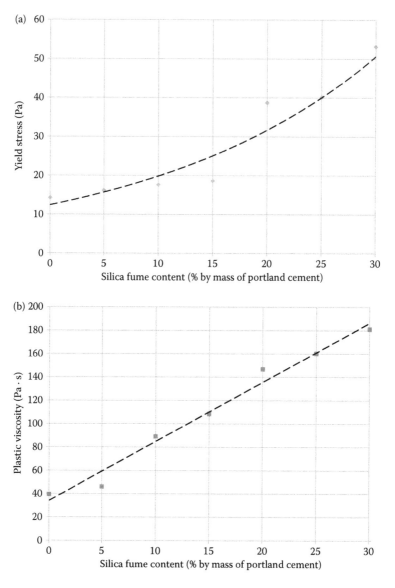

Figure 14.17 Effect of SF content on (a) yield stress and (b) plastic viscosity of SCCs. (Data from Benaicha, M. et al. 2015. *Construction and Building Materials*, 84, 103–110.)

that had been used for thousands of years in making adobe bricks: incorporating straw in the mud to reduce cracking due to shrinkage upon drying and to help holding the block together in dried state.

Fibre-reinforced concrete (FRC) is concrete with dispersed, randomly oriented and discrete fibres. Synthetic fibres such as steel, glass, plastic and

Table 14.3 Typical properties of various fibres used in FRCs

Fibre	Specific gravity	Modulus of elasticity (GPa)	Tensile strength (MPa)	Maximum elongation (%)
Acrylic	1.2	2.1	205–410	7.5–50.0
Bamboo	1.5	35	350–510	N/A
Carbon	1.6–2.1	30–480	480–4000	0.5–2.4
Glass	2.6	70–80	1000–3500	2.0–3.5
Jute	1.03	25–30	250–350	1.5–2.0
Polypropylene	0.9	3.5–5.0	135–700	15.0
Rock wool	2.7	70–120	480–750	0.6
Steel	7.8	200	275–2750	0.5–3.5

Source: Adapted from ACI 544. 1996. State-of-the-art report on fiber-reinforced concrete, ACI Committee 544 Report, ACI 544.1R-96, American Concrete Institute, Farmington Hills, MI; Mehta, P.K., Monteiro, P.J.M. 2006. *Concrete*, 3rd Ed., McGraw-Hill, New York; Mindess, S., Young, J.F. 1981. *Concrete*, Prentice-Hall, Englewood Cliffs, NJ.

carbon fibres and natural vegetable fibres may be used in producing FRC. Fiber type, geometry and amount affect the properties of FRC. Typical properties of various fibers are listed in Table 14.3.

FRCs are classified according to their fibre volume fractions as (a) low volume fraction, (b) moderate volume fraction and (c) high volume fraction. Low volume fraction (<1%) FRCs are mainly used to reduce shrinkage cracking; moderate volume fraction (1%–2%) FRCs have higher modulus of rupture and impact resistance; and high volume fraction (>2%) FRCs have much better performance in hardened state due to their strain-hardening response and they are called high-performance FRCs (Mehta and Monteiro, 2006). Although the latter FRCs usually require fibre volume fractions of around 10% or even more, recent advances in micromechanics and fibre technology enabled the achievement of high-performance FRCs with much lower fibre volume fractions (Li, 2002).

In most of the FRC applications, fibre volumes do not exceed 2%. Two main reasons for this are the increase in cost and the decrease in workability of concrete upon fibre incorporation. Although use of FRC is still very much limited to applications like industrial floors, tunnel linings, pavement overlays, pavement repairs etc. the vast amount of research carried out indicate that ductility, impact resistance, strength and first crack strength of concrete may be improved by proper use of fibres (Li, 2002).

Brittle materials like concrete break almost immediately after the initiation of the first crack which is defined as the point on the load-deflection curve of a concrete beam specimen at which the curve becomes nonlinear. However, FRC continues to carry loads after the first crack and results in increased work for fracture. Fibres may contribute to the energy absorption capacity and control of crack growth of FRC in several different ways. When the bond between the fibres and the matrix is weak, fibre pull-out occurs. If the bond is strong, either the fibres break or combined debonding of fibre and matrix would occur. In either of these two cases, the contribution of fibres to the mechanical properties is small. Therefore, the bond between the fibre and the matrix should be such that the fibres will act as crack bridges (Zollo, 1997; Mehta and Monteiro, 2006).

Most of the research reported on FRCs that contain mineral admixtures, unfortunately do not concentrate on the effect of mineral admixtures. Some are related with the general properties such as strength and durability but not with energy absorption capacity. Use of mineral admixtures in FRCs seems to result in similar changes in fresh and hardened properties as they do in normal concrete.

Impact resistance at failure of polypropylene FRC was found to increase 82%, 42% and 90% by partially replacing PC with 25% FA, 10% SF and 25% GGBFS in PC, respectively. Increased effectiveness of the fibres in the presence of mineral admixtures was attributed to the improved bonding which was associated with the pozzolanic action (Alhozaimy et al., 1996).

Similarly, incorporation of 8% (by mass) SF was found to improve the energy absorption capacity of steel FRCs considerably. The drop-weight test described by ACI 544 (1989) was used on two series of steel FRCs with water–cementitious ratios of 0.45 and 0.36. 0.5% and 1% (by volume) steel fibres were used in each series. Number of blows for first crack increased 1.5 and 2.4 times in 0.45 water–cementitious ratio FRCs with 0.5% and 1.0% steel fibres, respectively. Increases were 1.3 and 1.5 times for 0.36 water–cementitious ratio FRCs. Number of blows for failure were recorded to increase 1.2, 2.2 and 1.3, 1.5 times, respectively (Nili and Afroughsabet, 2010).

A new class of fibre-reinforced cementitious composites with very high ductility and toughness which is attained by a sort of strain-hardening mechanism has attracted much attention by researchers. These materials are called strain-hardening cementitious composites (SHCC) or more commonly nowadays, engineered cementitious composites (ECC). The mechanical properties of ECC depend on the type and amount of constituents used in the mix and particularly on the characteristics of the fibres. High ductility is a result of fibre-matrix-interface interaction. However, besides the individual influence of each parameter, their combined influence should be considered through composite optimisation (Li, 2002).

In a typical ECC, only fine sand with maximum aggregate size of about 250 μm is used as aggregate and the cement content may be more than 1000 kg/m³. Thus, many investigations (Kim et al., 2007; Zhu et al., 2012; Zhang et al., 2014; Zhu et al. 2014) on incorporating mineral admixtures in ECC to partially replace cement were carried out. Consequently, mineral admixtures such as FA and GGBFS are now being considered as essential cementitious materials in producing ECC and their amount may even be higher than that of PC in the mixture. Furthermore, pozzolanic mineral admixtures were also found to improve many durability properties and reduce shrinkage of ECCs (Şahmaran and Li, 2008; Şahmaran et al., 2009, 2013).

The partial replacement of PC by mineral admixtures is also a common practice in slurry infiltrated fibre concrete (SIFCON). High volume fractions (5%–30%) of fibres are placed in the forms and a high water–cement ratio and cement rich mortar slurry is infiltrated to cover the fibres to produce SIFCON. Using a high amount of PC in SIFCON, which not only affects the cost but also the heat of hydration and shrinkage, may be disadvantageous. Therefore, mineral admixtures are used to solve these problems and to also improve the durability characteristics (Yazıcı et al., 2006).

14.6 REACTIVE POWDER CONCRETE

Reactive powder concrete (RPC) is a cementitious material which is characterised by very high strength and very low porosity which are basically achieved by optimising the particle packing and using low water contents. It does not contain coarse aggregate. Fine sand, crushed quartz or sometimes SF with particle sizes ranging between 0.02–300 µm are the only aggregates in this composite material. A high amount of water-reducing agent is necessary to keep the water–cementitious ratio low. Usually, the water–cementitious ratio is less than 0.20. The particle-size distribution is optimised to improve the packing and to densify the mixture. Further improvement of the microstructure may be achieved by curing the material at high temperatures and/or applying pressure and ductility may be increased by incorporating short fibres (Yazıcı et al., 2009).

Mix proportions of RPCs from various selected investigations are given in Table 14.4. As seen from the table and in many other investigations (Richard and Cheyrezy, 1995; Yunsheng et al., 2008; Kumar, et al., 2013) SF is an essential constituent of RPC. Besides SF, fly ashes and GGBFS were successfully used to partially replace PC in producing RPCs with satisfactory properties.

Table 14.4 Mix proportions of RPC in various selected studies

Constituents (kg/m³)	Yazıcı et al., 2009		Yiğiter et al., 2012	Zenati et al., 2009
Cement	830	581	752	750
SF	291	157	282	44
GGBFS	–	83	–	88
FA	–	166	188	–
Aggregate				1235 (Total)
1–3 mm	489	534	875	
0.5–1 mm	244	266	224 (0–1 mm)	
0–0.4 mm	244	266		
Water	151	151		212
HRWRA	55	34		13.5

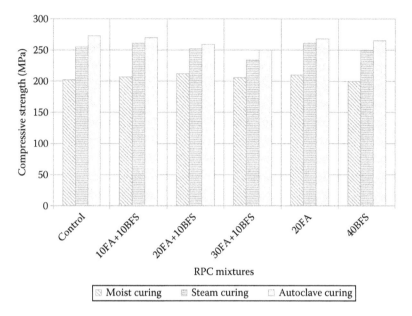

Figure 14.18 Compressive strength of high-lime fly ash and GGBFS-incorporated RPCs under different curing regimes. (Data from Yazıcı, H. et al. 2009. *Construction and Building Materials*, 23, 1223–1231.)

Yazıcı et al. (2009) used a high lime fly ash and a GGBFS to replace 20%–40% of PC in binary and ternary mixtures. All mixtures had an SF-to-PC ratio of 0.20–0.25 (by mass). RPCs were cured in water for 28 days, under atmospheric pressure steam at 100°C for 3 days, and in autoclave under 2 MPa pressure and at 210°C for 8 h. All RPCs had compressive strengths above 200 MPa and incorporating FA and or GGBFS did not affect the strength, as shown in Figure 14.18.

Yunsheng et al. (2008) designed three RPCs with matrices composed of (1) 50% PC + 25% FA + 25% GGBFS, (2) 50% PC + 10% SF + 40% GGBFS and (3) 40% PC + 10% SF + 25% FA + 25% GGBFS. Three types of curing regimes as standard moist curing, steam curing (90°C) and autoclave curing (1.7 MPa pressure and 200°C) were applied to the RPCs prepared. They were able to get compressive strengths between 120 and 140 MPa under standard curing conditions and considerably higher strengths under steam and autoclave curing. Among the three matrices investigated, highest strengths were obtained in the third mix which contained the least amount of PC.

14.7 MASS CONCRETE

Cement and Concrete Terminology reported by ACI Committee 116, defines mass concrete as 'any volume of concrete with dimensions large enough to require that measures be taken to cope with generation of heat from hydration of the cement and attendant volume change, to minimize cracking' (ACI 116, 2000). This definition implicitly states that the term mass concrete should not only be associated with large concrete dams as it is usually done. In fact, problems related with heat generation may be encountered in concrete elements with the smallest cross sectional dimension larger than 60 cm, if proper precautions are not taken.

Hydration of cement is an exothermic process that leads to temperature rise at the core portion of a large mass of concrete. Heat generated will not be easily dissipated since concrete is a fair insulator. The temperature increase would lead to tensile stresses which may easily exceed the tensile strength of concrete. On the other hand, the surface of the concrete would easily be cooled off resulting in contraction. The temperature gradient may cause cracking of concrete by thermal expansion and contraction. Thus, the control of heat generation becomes the primary concern in mass concreting which can be achieved by means of proper selection of materials, mix proportions and construction practices. Most specifications on mass concrete require that 7- and 28-day heat of hydration of cement should be at the highest 250 kJ/kg and 290 kJ/kg, respectively. Considering that the average heat of hydration values of ordinary PCs are about 370 kJ/kg and 440 kJ/kg at 7- and 28-days, it is obvious that the requirement stated above cannot be met with ordinary PCs. ASTM C 150 (2012) specifies a moderate-heat Type II and a low-heat Type IV PCs with maximum 7-day heat of hydration values of 290 kJ/kg and 250 kJ/kg, respectively. The former is not able to fulfill the heat generation requirement in mass concretes and the latter is not being commonly produced any more because there are other more economical ways of having low heat generation. One of the most common ways of controlling temperature rise in mass concrete is to use pozzolans and GGBFS as cement replacement materials. The first recorded use of pozzolan in mass concrete dates back to 1920s. 20% (by mass) PC was replaced by pumicite in the concrete of an abutment of Big Dalton Dam in Los Angeles, USA (ACI 207, 2004). Later on, use of pozzolans or GGBFS at a rate of 30% (by mass) or more in mass concreting had become an almost regular practice.

Results of an experimental investigation on the influence of a natural pozzolan and a GGBFS when used to replace 20% and 35% PC on the 7- and 28-day heat of hydration are shown in Figure 14.19. PC and mineral admixtures were separately ground to 300 ± 100 m²/kg Blaine fineness and then blended.

14.8 ROLLER COMPACTED CONCRETE

Roller compacted concrete (RCC) emerged as a new type of mass concrete in late 1950s. Its first recorded use was in the Shihmen cofferdam in Taiwan (Özcan, 2008). Currently, there are over 650 RCC dams built throughout

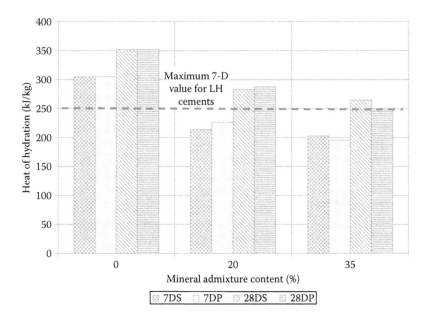

Figure 14.19 Effect of natural pozzolan and GGBFS on heat of hydration. (Data from Tokyay, M., Delibaş, T., Aslan, Ö. 2010. *Effects of mineral admixture type, Grinding Process, and Cement Fineness on the Physical and Mechanical Properties of GGBFS-, Natural Pozzolan-, and Limestone-incorporated Cements,* Working Paper AR-GE 2010/01-B, Turkish Cement Manufacturers' Association [TÇMB], 45p. [in Turkish].)

the world (de Aldecoa and Ortega, 2013; Lewis, 2015). Use of RCC is not limited to dam construction but also to concrete pavements (ACI 325, 2001).

RCC is a no-slump concrete which is able to support a vibratory roller during compaction (ACI 207, 1999). Mix proportioning of RCC is based on similar considerations as mass concrete: basically, coarse aggregate content is maximised and the cement content is minimised for the specified workability, strength, permeability and heat generation requirements. There are several different mix proportioning methods that fall under two general categories as (a) 'concrete' approach and (b) 'soils' approach. This book is not intended to deal with RCC in detail. Therefore, the reader is referred to ACI 211.3 (2002) for these methods.

RCC can be made with any of the basic types of PCs. However, in mass concrete applications, portland-pozzolan or portland-BFS cements or combinations of pozzolans or GGBFS with PCs of moderate heat of hydration should be preferred. Use of pozzolans and GGBFS in RCC may serve other purposes such as providing additional fines content for improving the workability, increasing the paste volume and reducing the cost, besides lowering heat generation. Pozzolan content in RCC may be as high as 80%, by mass of total cementitious material (ACI 207, 1999).

Chapter 15

Mineral admixtures as primary components of special cements

15.1 SUSTAINABILITY AND CEMENT

Portland-based cements are the most common inorganic binders used in concrete manufacturing. According to the European Cement Association, world cement production in 2014 was estimated as 4.3 billion tons (CEMBUREAU, 2015). When compared with figures at the turn of twentieth century, there appears to be more than 130% increase in cement production which indicates that portland-based cements will continue to be key construction materials in the foreseeable future. On the other hand, the cement industry is confronted with serious energy and emissions problems. Cement manufacturing accounts for approximately 75% of total energy use in non-metallic minerals production and about 25% of total industrial CO_2 emissions (IEA, 2007). These values correspond to about 2% and 5% of total global and total industrial energy consumption, respectively (Placet and Fowler, 2002). Energy represents about 30% of the total cost of cement and the main energy consuming process in cement manufacturing is the burning process in the rotary kiln. A typical modern rotary kiln consumes 3.1 GJ/ton of clinker produced. The value may be as high as 6.5 GJ/ton in less efficient kilns and for wet raw materials. The world average is reported as 3.8 GJ/ton. Besides, total electrical energy consumption is stated to be around 0.38 GJ/ton. Carbon dioxide (CO_2) emission due to carbon-based fuels is approximately 0.3 kg/kg of clinker in a modern rotary kiln and may increase up to 0.6 kg/kg of clinker in less efficient kilns. Besides the fuel derived CO_2, about 0.57 kg/kg of clinker CO_2 is given off through the calcination of raw materials, basically limestone (Gartner, 2004; CSI, 2005; Damtoft et al., 2008). Therefore, there has been a continuous effort in increasing the energy efficiency and decreasing CO_2 emissions in cement manufacturing. In fact, many improvements in process efficiency have been achieved by the cement industry. However, emissions are continuing to increase globally due to the increase in population and growing industrialisation especially in developing countries (Imbabi et al., 2012).

Currently, there are five main routes proposed for reducing the carbon footprint of cement: (1) increasing resource efficiency, (2) increasing energy

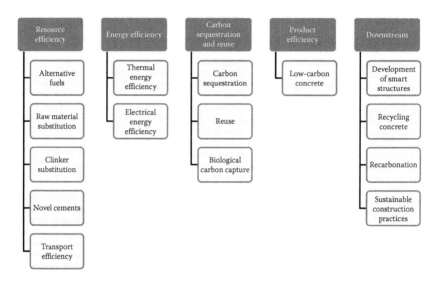

Figure 15.1 Five parallel routes and related practices for reducing the carbon foot-print of cement. (Adapted from CEMBUREAU. 2013. The Role of Cement in the 2050 Low-Carbon Economy, CEMBUREAU The European Cement Association, Brussels, 64pp.)

efficiency, (3) carbon sequestration and reuse, (4) product efficiency and (5) downstream measures. The first three are directly related with the cement sector and the last two are related with the construction sector. A potential reduction of carbon footprint up to 80% was stated to be achievable if these five parallel routes are properly taken (CEMBUREAU, 2013). The five routes and related practices are shown in Figure 15.1. Although most of the practices listed in the figure are beyond the scope of this book, each will be very briefly described.

15.1.1 Resource efficiency

Alternative fuels including a high amount of waste products such as biomass, pretreated sewage sludge, tyres, waste oil and solvents and saw dust may be used in cement production. Besides partially replacing carbon-intensive fossil fuels, their use may also help in reducing landfill problems and their own greenhouse gas emissions when incinerated.

About 65% of CO_2 emissions of the cement industry result from the decarbonation of raw materials during the burning process. The majority of it comes from the dissociation of limestone. Various selected waste and by-products that contain lime, silica, alumina and iron oxide may be used to partially replace the raw materials. BFS, FA, demolition wastes and ashes of various alternative fuels are such materials that may be used.

Reduction of clinker-cement ratio by blending PC with alternative materials like pozzolans, limestone powder and latent hydraulic materials had already become a common practice.

Several types of novel, low-energy and low-CO_2 clinkers and cements are being developed. Their uses are still very limited due to difficulties related with validation, market acceptance, availability and non-existence of standards. Nevertheless, the attempts for such cements are exciting, although some of them remain to be of scientific interest, at least for the time being.

Instead of road transportation, rail and water transportation of cement over long distances would reduce transport emissions as well as reducing the cost.

15.1.2 Energy efficiency

Modernising cement plants, replacing older plants, using dry process with preheating and precalcination technology, waste heat recovery systems and modern grinding technology would improve both thermal and electrical energy efficiency.

15.1.3 Carbon sequestration and reuse

Besides the procedures and methods briefly described in the previous sections, a technology which is in the research and development stage may provide the cement industry with much less CO_2 emissions: CO_2 capture-storage or reuse (CCS or CCU). CO_2-capture may be done by means of chemical absorption, membrane technologies and carbonate looping (reforming calcium carbonate). However, there are still many unresolved points related with CCS and CCU from the economical, ecological and legal points of view. CCS may be realistic if CO_2 transportation and storage places are suitable, approved and accepted by society.

There is also ongoing research for biological carbon capture. In very simple terms, it is thought that CO_2 from the stack is absorbed by algae to be used for photosynthesis. Then the grown algae is harvested, dried and used as fuel. The method is in its very early development state (CEMBUREAU, 2013).

15.1.4 Product efficiency and downstream measures

Although concrete may be considered a low-carbon material, using high performance concretes, optimisation of cement, mineral and chemical admixture contents and using locally available aggregates would result in further slight decreases in the carbon footprint.

Buildings that require less energy with increased thermal efficiency, new and efficient methods of recycling and deconstruction of concrete, taking measures to increase the service life of concrete, making use of concrete's thermal mass for energy efficiency and many more innovative techniques may be applied.

15.2 LOW-ENERGY, LOW-CO$_2$ CEMENTS WITH MINERAL ADMIXTURES

Use of mineral admixtures, besides PC clinker, as a main constituent in blended cements or as partial replacement for PC in concrete had already become a common practice. Besides their technical and economic advantages, they also have an undisputed role from the ecological point of view.

Many of the technical influences of mineral admixtures when used as ingredients of cement and concrete were covered so far. Their influences on energy requirement and CO$_2$ emissions are discussed in this chapter.

15.2.1 Mineral admixtures as clinker substitutes

Using mineral admixtures as clinker substitutes reduces both energy consumption and CO$_2$ emissions. Calcination and grinding are the two processes that result in the highest share of energy requirement. Most of the mineral admixtures, with the exception of granulated BFS, are either softer than clinker or may even require no grinding due to their already fine particle sizes. Therefore, the grinding energy may be reduced by clinker substitutes. Since no calcination is necessary for most of the mineral admixtures, the thermal energy requirement will also be reduced. Globally, clinker-to-cement ratio in 2003 was estimated as 0.85 (Damtoft et al., 2008) which means that about 15% non-clinker materials (including about 4%–5% gypsum) were present in the cements. There had been some gradual but steady reduction in clinker-to-cement ratio between 1994 and 2004 from about 0.87 to about 0.77 (IEA, 2007). However, there are various uncertainties with the data available and further use of mineral admixtures directly in concrete would affect the statistical figures. Nevertheless, considering the possibility of producing cements with clinker-to-cement ratios as low as 0.05 (as for CEM III/C in EN 197-1) and having satisfactory properties, it seems that mineral admixtures will be used in much higher amounts as clinker substitution. The beneficial effects of mineral admixtures on energy requirement and CO$_2$ emissions are directly related with the amount of these materials used to replace the clinker.

15.2.2 Mineral admixtures as raw materials for low-energy cements

There has been continuous research to produce cements with lower energy than that required for PCs. Although none of the low-energy or low-CO_2 cements developed so far are competitive alternatives of portland-based cements yet, some of them are worth mentioning: belitic cements, alinite cements, sulphoaluminate belite (SAB) cements and alkali-activated binders. Before discussing such novel cements and possibilities of using mineral admixtures as constituents, it should be restated that lime–pozzolan mixtures which can be considered as the first low-energy and low-CO_2 inorganic binder have been in use for thousands of years (Bensted and Coleman, 2003) and hydraulic lime obtained by the calcination of siliceous or argillaceous limestone at temperatures around 1300°C were used as cements long before the advent of PC (Grist et al., 2015).

15.2.2.1 Belitic cements

Alite (C_3S) and belite (C_2S) are the two most important compounds of a PC. The total energy required for their formations are 1770 and 1268 kJ/kg, respectively (Odler, 2000). Besides, C_3S synthesis needs more CaO than C_2S synthesis which means a higher amount of $CaCO_3$ in the raw meal is necessary for C_3S, leading to more CO_2 emissions, as given by

$$\underbrace{CaCO_3 \rightarrow CaO + CO_2}_{\text{Decomposition of limestone}} \tag{15.1}$$

$$\underbrace{2CaO + SiO_2 \rightarrow Ca_2SiO_4}_{C_2S \text{ formation}} \tag{15.2}$$

$$\underbrace{3CaO + SiO_2 \rightarrow Ca_3SiO_5}_{C_3S \text{ formation}} \tag{15.3}$$

Therefore, one of the ways of reducing the energy consumption and CO_2 release during cement production is to increase C_2S content and decrease C_3S content which can be achieved by using raw meal with a lower lime saturation factor (LSF). When LSF is reduced from 100 to 75, there results a 12% decrease in heat requirement and 6% decrease in CO_2 emission. Such a clinker does not contain C_3S. However, C_2S is not as reactive as C_3S and the early strength loss in PCs with LSF < 80% cannot be compensated even with very fine grinding. On the other hand, it was suggested that the reactivity of belitic cements may be increased by rapid quenching of the clinker or by substituting alkali or sulphate ions into the crystal structure during formation (Lawrence, 1988).

Reactive belite cements were obtained under laboratory conditions by using different techniques such as mechanical activation (very fine grinding), hydrothermal pretreatment at 100°–275°C of lime and silica blends and then calcination at 850°–1300°C, or drying colloidal silica–calcium nitrate solution at 70°C and then calcining at 760°C (Chatterjee, 1996a,b; Odler, 2000).

Similar methods were employed for producing belitic cement of satisfactory performance by using fly ashes or slags. Guerrero et al. (1999) blended CaO with a low-lime fly ash at a Ca–Si molar ratio of 2.0. The blend was mixed with water at a water–solids ratio of 5.0. The mixture was treated at 200°C, 1.24 MPa pressure for 4 h. Then the dried sample was calcined at 700°C, 800°C and 900°C. The main phases obtained were α'-C_2S, β-C_2S and $C_{12}A_7$. The fastest and highest hydraulic activity was observed in a sample calcined at 800°C. Later, Guerrero et al. (2004) tried a high-lime fly ash by itself, hydrothermally treated with 1M NaOH solution at 100°C, 150°C and 200°C and calcined at 700°C, 800°C, 900°C and 1000°C afterwards. They were able to get α-C_2S with high-lime fly ash, alone.

Pimraksa et al. (2009) synthesised belite cements from the mixtures of FA and $CaCO_3$ and RHA and $CaCO_3$ at Ca–Si ratios of 2.0–3.0. In one of the series, $BaCl_2$ and gypsum were also used in different amounts. Each of the mixtures was combined with NaOH solution and pretreated at 130°C and 1 kg/cm^2 pressure for 4 h. Then calcination was carried out at 750°C, 850°C and 950°C. Upon hydration, C-S-H and C-A-H were obtained. They also determined the strength of the cements to be between 5.0 and 9.5 MPa at 28 days.

In another study, belitic cements were obtained by incorporating 5% and 10% (by mass) electric arc furnace slag into limestone–clay mixtures with LSF = 78.0 and 80.0. The raw meals were calcined at 1380°C and then fast cooled. The resulting clinkers contained C_2S, C_3S, C_3A and C_4AF (Iacobescu et al., 2011).

15.2.2.2 Alinite cements

Alinite is a calcium oxy-chloro-aluminosilicate with the following formula:

$$Ca_{10}Mg_{1-\frac{x}{2}}\square_{\frac{x}{2}}\left[(SiO_4)_{3+x}(AlO_4)_{1-x}O_2Cl\right] \quad 0.35 < x < 0.45 \quad (15.4)$$

where, $\square$ refers to a lattice vacancy (Güneş, 2010).

Alinite may be synthesised using a mix composed of $CaCO_3$, SiO_2, Al_2O_3, MgO and $CaCl_2$ which is burned at 1000°–1300°C. The structure of alinite is similar to alite but oxygen and silica are partially substituted by chlorine and alumina, respectively. The crystal lattice of alinite may accommodate limited amounts of ions such as Fe^{3+}, P^{5+}, Ti^{4+}, Na^+ and K^+ (Odler, 2000; Pradip and Kapur, 2004).

Alinite cement clinkers are produced from raw meals containing limestone, clay, MgO and $CaCl_2$. The amount of calcium chloride in the raw mix is 6%–18%. Clinkerisation temperature is around 1150°–1200°C. When compared with PC clinker, alinite cement clinker would result in upto 30% energy saving (Güneş, 2010). Alinite cement clinkers are always soft and friable due to weak Ca–Cl bonds. Therefore, grinding energy requirement is also less for these clinkers (Pradip and Kapur, 2004).

The amount of limestone necessary to produce alinite cement clinker is around 60%–70% which is less than that is generally required for a PC clinker. Therefore, amount of CO_2 given off is less in alinite cement production.

Hydration of alinite cement results in C-S-H and CH form as in PC hydration. Besides, phases like Friedel's salt, monosulphate hydrate (AFm), ettringite (Aft) and small amounts of CAH_{10} and C_3AH_6 may be present.

Alinite cements have sufficient strength properties which are comparable to or even sometimes higher than those of ordinary PCs (Odler, 2000; Pradip and Kapur, 2004; Güneş, 2010). Gypsum addition to the final product is reported to increase the strength proportional to the amount of gypsum added (Odler, 2000).

Alinite cements have been produced since the 1970s. There are several commercial alinite cement manufacturers in countries like India, Russia and Japan (Gür et al., 2010). However, it should be noted that chlorine corrosion of the production equipment needs to be taken into account in alinite cement manufacture (Odler, 2000).

Pradip and Kapur (2004) produced three alinite cements from mixtures of (1) FA, limestone fines, mill scale (Fe_2O_3) and magnesite ($MgCO_3$) dust, (2) gold mine tailings, limestone and waste hydrochloric acid and/or calcium chloride as flux, (3) municipal incinerator ash, limestone and siliceous sand. They determined that the lime index (LI) which is defined as the ratio of CaO to ($SiO_2 + Al_2O_3 + Fe_2O_3$) of the raw mix should be above 1.5 and optimum chloride content is around 8%, by mass. The amount of gypsum added was around 8%. They were able to get compressive strength of 14.5–26.0 MPa, 16.0–47.5 MPa and 23.5–50.0 MPa at 3, 7 and 28 days, respectively. Besides, they estimated 30%–60% thermal energy and 50%–80% electrical energy saving when compared with PC.

Municipal incinerator ash (MIA) was successfully used in another study to produce alinite cement (Singh et al., 2008; Wu et al., 2012). Good quality alinite clinkers were obtained with mixes having MIA contents 25%–45% and calcined at 1100°–1200°C for 2 h. Alinite cements were made by blending the ground clinker with 5%–8% gypsum.

Güneş et al. (2012) produced alinite cement by mixing dried soda sludge (73.5%), clay (26.3%) and iron ore (0.2%). Soda sludge is a waste of the Solvay process that has 30%–60% water. A typical dried soda sludge contains around 50% $CaCO_3$, 12% $CaSO_4$, 12% $Mg(OH)_2$, 9% $Ca(OH)_2$, 5% $CaCl_2$ and 3% $NaCl_2$ (Shatov et al., 2004). The clinkers were produced at burning temperatures of 1050°–1200°C. After adding 5% (by mass) gypsum, 28 days compressive strength of 26.6 MPa was attained. It was concluded that soda sludge can be used to obtain low-energy alinite cements which do not require any natural limestone.

15.2.2.3 SAB cements

SAB cements contain two basic phases, belite (C_2S) and calcium sulphoaluminate ($C_4A_3\bar{S}$). Generally, they constitute 25%–65% and 10%–20% of the SAB cement, respectively. C_4AF may also be present in significant amounts, sometimes as high as 40%. Some minor phases like CA, $C_{12}A_7$, C_2AS and $C_5S_2\bar{S}$ may also appear. A suitable raw mix of appropriate oxide composition consisting of limestone, clay, bauxite and gypsum is burnt at 1200°–1300°C to produce SAB cements. The calcination energy requirement and CO_2 emision of SAB cement are lower not only due to lower burning temperature but also due to the lower CaO content of the clinker. On the other hand, since gypsum is part of the raw mix, additional measures may be necessary for SO_2 emissions.

The main hydration products of SAB cement are C-S-H gel and ettringite. Ettringite forms readily upon hydration and is responsible for early strength whereas C_2S hydration results in C-S-H gel at ages beyond 7 days and is responsible for late strength. SO_3-Al_2O_3 molar ratio was suggested to be between 1.3 and 1.9 for optimising the strength (Odler, 2000). A recent study on synthesising sulphoaluminate cement from a high alumina FA and a flue gas desulphurisation (FGD) gypsum as raw materials showed that 3-, 7- and 28-days strength of high belite sulphoaluminate cements increase with the increasing amount of FGD gypsum, upto 15% (by mass). 20% gypsum results in slight decrease in strength (Ma et al., 2013). The strength results shown in Figure 15.2 were obtained on mortar specimens with a water–cement ratio of 0.38.

SAB cements from FA-lime kiln baghouse dust-scrubber sludge mixtures were obtained at 1175°C calcination temperature for 1 h. Isothermal

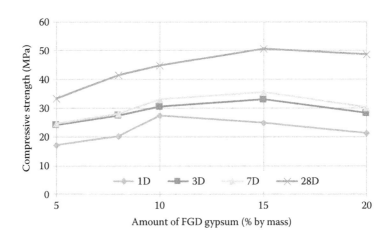

Figure 15.2 Compressive strength of high belite sulphoaluminate cements with varied amount of FGD gypsum. (Data from Ma, B. et al. 2013, *Ceramics-Silikáty*, 57(1), 7–13.)

calorimetry and microstructural studies and strength tests have shown that the properties of the SAB cement were comparable to ordinary PC (Arjunan et al., 1999).

Several other industrial waste or by-products such as red mud, FA, BFS, phosphogypsum, desulphogypsum, aluminous by-products of secondary aluminium manufacture, bottom ash and electric arc furnace slag were used as raw materials partially replacing or substituting the natural raw materials in producing SAB cements (Duvallet et al., 2009; Marroccoli et al., 2010; Ukrainczyk et al., 2013).

15.2.2.4 Alkali-activated binders

Alkali-activated binders are made by mixing powdered aluminosilicate materials with alkaline activating solutions. Although there are activators like alkali sulphates, alkali carbonates and alkali aluminates reported (Pacheco-Torgal et al., 2008) common activators used are sodium hydroxide (NaOH) and sodium silicate (Na_2SiO_3). The chemistry of the alkali activation is not yet fully understood (Pacheco-Torgal et al., 2008; Juenger et al., 2011). Reaction mechanisms involved in alkali activation depend both on the nature of the aluminosilicate material and the activator. The reaction products may vary depending on the level of calcium that is available for reaction. A high-calcium system usually results in amorphous or partially crystalline C-(A)-S-H gel with low lime–silica ratio whereas a low-calcium system results in N-A-S-H gel with a zeolite-like structure as the primary reaction products. Secondary phases like hydrotalcite, strätlingite, xonotlite, hydroxysodalite etc. may also appear (Pacheco-Torgal et al., 2008; Juenger et al., 2011).

Alkali-activated binders have received much attention because of their possibly low environmental impact. They can be produced almost completely from waste or by-product materials such as BFSs and fly ashes without any need for a calcination process. Among the vast number of publications on alkali-activated slag, FA or other similar aluminosiliceous materials, a few recent studies are summarised here.

Altan and Erdoğan (2012) investigated the strength development of alkali hydroxide and sodium silicate solution activated slag mortars with different water-slag ratios cured at 5°C, 23°C and 80°C. Although, specimens cured at 80°C achieved strengths above 70 MPa within the first day, specimens cured at room temperature gained comparable or even higher strengths when sufficient time is given. Specimens cured at low temperature (5°C) reached about 20 MPa compressive strength in 28 days. KOH was found to be a more effective activator than NaOH at high temperatures. At room temperature however, although KOH resulted in higher strengths in the first few weeks, NaOH-activation caused higher strengths at later ages. Sodium silicate was found essential only for high temperature curing condition.

Aydın and Baradan (2014) studied the effect of SiO_2–Na_2O ratio (M_s) of NaOH–Na_2SO_3 solutions on BFS activation. M_s values and Na_2O concentrations of the activator varied from 0.4 to 1.6 and 2% to 8%, respectively. They also determined various fresh and hardened properties of alkali-activated slag mortars. All specimens were moist cured at 20°C. For high Na_2O concentrations and low M_s values, the mortars were not sufficiently workable. 2-days strength of alkali-activated slag mortars were lower than the control PC mortars. However, for $M_s \geq 0.8$ and $Na_2O \geq 4\%$, equivalent or higher strengths were obtained at and beyond 7 days.

A low-lime fly ash was activated by using sodium hydroxide and sodium silicate solutions by Zhang et al. (2014). The 90-days compressive strength achieved at 23°C curing when a 12 M NaOH was used as activator was 4.0 MPa. Whereas, using $Na_2O \cdot 1.5SiO_2$ as activator, 53.2 MPa strength was obtained. By using 80:20 mixture of FA and slag and $Na_2O \cdot 1.5SiO_2$ as activator, they were able to attain 77.4 MPa strength at 90 days.

A 25% FA and 75% BFS blend was activated with a solution having $M_s = 1.0$ and 10% Na_2O concentration and cured at 95°C for 24 h to attain about 80 MPa compressive strength (Marjanović et al., 2015).

References

Abdelkader, B., El-Hadj, K., Karim, E. 2010. Efficiency of granulated blast furnace slag replacement of cement according to the equivalent binder concept, *Cement and Concrete Composites*, 32, 226–231.

Abdulmaula, S., Odler, I. 1981. Hydration reactions in fly ash-portland cements. In *Proceedings of the Symposium N on Effects of Fly Ash Incorporation in Cement and Concrete* (Ed. V.M. Malhotra), Materials Research Society, Boston, USA, pp. 102–111.

Abrams, D.A. 1919. *Design of Concrete Mixtures*, Bulletin 1, Structural Materials Research Laboratory, Lewis Institute, Chicago, 20pp.

Achtemichuk, S., Hubbard, J., Sluce, R., Shehata, M.H. 2009. The utilization of recycled concrete aggregate to produce controlled low-strength materials without using portland cement, *Cement and Concrete Composites*, 31, 564–569.

ACI 116. 2005. Cement and concrete terminology, ACI Committee 116 Report, ACI 116R-00, American Concrete Institute, Farmington Hills, MI.

ACI 201. 2001. Guide to durable concrete, ACI Committee 201 Report, ACI 201.2R01, American Concrete Institute, Farmington Hills, MI.

ACI 207. 1999. Roller-compacted mass concrete, ACI Committee 207 Report, ACI 207.5R-99, American Concrete Institute, Farmington Hills, MI.

ACI 207. 2004. Mass concrete, ACI Committee 207 Report, ACI 207.1R-96, American Concrete Institute, Farmington Hills, MI.

ACI 209. 1997. Prediction of creep, shrinkage, and temperature effects in concrete structures, ACI Committe 209 Report, American Concrete Institute, Farmington Hills, MI.

ACI 211. 1997. Standard practice for selecting proportions for normal, heavyweight, and mass concrete, ACI Committee 211 Report, ACI 211.1–87, American Concrete Institute, Farmington Hills, MI.

ACI 211.3. 2002. Guide for selecting proportions for no-slump concrete, ACI Committee 211 Report, ACI 211.3R-02, American Concrete Institute, Farmington Hills, MI.

ACI 212. 2010. Report on chemical admixtures for concrete, ACI Committee 212 Report, ACI 212.3R-04, American Concrete Institute, Farmington Hills, MI.

ACI 213R. 2014. Guide for structural lightweight aggregate concrete, ACI Committee 213 Report, ACI 213R-14. American Concrete Institute, Farmington Hills, MI.

ACI 221. 1998. State-of-the-art report on alkali–aggregate reactivity. ACI Committee 221 Report, ACI 221.1R-98, American Concrete Institute, Farmington Hills, MI.

ACI 229. 2005. Controlled low-strength materials, ACI Committee 229 Report, ACI 229R-99, American Concrete Institute, Farmington Hills, MI.

ACI 232.1R. 2001. Use of raw or processed natural pozzolans in concrete, ACI Committee 232 Report, ACI 232.1R-00, American Concrete Institute, Farmington Hills, MI.

ACI 232.2R. 2003. Use of fly ash in concrete. ACI Committee 232 Report, ACI 232.2R-03, American Concrete Institute, Farmington Hills, MI.

ACI 233. 2003. Slag cement in concrete and mortar. ACI Committee 233 Report, ACI 233R-03, American Concrete Institute, Farmington Hills, MI.

ACI 234. 2000. Guide for the use of silica fume in concrete. ACI Committe 234 Report, ACI 234R-96, American Concrete Institute, Farmington Hills, MI.

ACI 237. 2007. Self-consolidating concrete. ACI Committee 237 Report, ACI 237R-07, American Concrete Institute, Farmington Hills, MI.

ACI 308. 2001. Guide to curing concrete. ACI Committe 308 Report, ACI 308R-01, American Concrete Institute, Farmington Hills, MI.

ACI 318. 2008. Building code requirements for structural concrete. ACI Committee 318 Report, American Concrete Institute, Farmington Hills, MI.

ACI 325. 2001. Report on roller-compacted concrete pavements. ACI Committee 325 Report, ACI 325.10R-95, American Concrete Institute, Farmington Hills, MI.

ACI 363. 1997. State-of-the-art report on high-strength concrete. ACI Committee 318 Report, American Concrete Institute, Farmington Hills, MI.

ACI 544. 1989. Measurement of properties of fiber-reinforced concrete. ACI Committee 544 Report, ACI 544.2R-89, American Concrete Institute, Farmington Hills, MI.

ACI 544. 1996. State-of-the-art report on fiber-reinforced concrete, ACI Committee 544 Report, ACI 544.1R-96, American Concrete Institute, Farmington Hills, MI.

Adamapoulou, E., Pipilikaki, P., Katsiotis, M.S., Chaniotakis, M., Katsioti, M. 2011. Hoe Sulfates and increased temperature affect delayed ettringite formation (DEF) in white cement mortars, *Construction and Building Materials*, 25, 3583–3590.

Adesanya, D.A., Raheem, A.A. 2009. Development of corn cob ash blended cement, *Construction and Building Materials*, 23, 347–352.

Ahari, R.S., Erdem, T.K., Ramyar, K. 2015. Permeability properties of self-consolidating concrete containing various supplementary cementitious materials, *Construction and Building Materials*, 79, 326–336.

Ahmad, S. 2003. Reinforcement corrosion in concrete structures, its monitoring and service life prediction – a review, *Cement and Concrete Composites*, 25, 459–471.

Aïtcin, P.-C. 1995. Developments in the application of high-performance concretes, *Construction and Building Materials*, 9(1), 13–17.

Aïtcin, P.C., Autefage, F., Carles-Gibergues, F., Vaquier, A. 1986. Comparative study of the cementitious properties of different fly ashes, *Proceedings of the 2nd International Conference on Fly Ash, Silica Fume, Slag, and Natural Pozzolans in Concrete*, Madrid, ACI SP-91, Vol. 1, pp. 91–114.

Aïtcin, P.C., Pinsonneault, P., Roy, D.M. 1984. Physical and chemical characterization of condensed silica fume, *Bulletin of the American Ceramic Society*, 63, 1487–1491.

Akçaoğlu, T., Tokyay, M., Çelik, T. 2002. Effect of coarse aggregate on interfacial cracking under uniaxial compression, *Materials Letters*, 57, 828–833.

Akçaoğlu, T., Tokyay, M., Çelik, T. 2004. Effect of coarse aggregate size and matrix quality on ITZ and failure behavior of concrete under uniaxial compression, *Cement and Concrete Composites*, 26, 633–638.

Alhozaimy, A.M., Soroushian, P., Mirza, F. 1996. Mechanical properties of polypropylene fiber-reinforced concrete and the effects of pozzolanic materials, *Cement and Concrete Composites*, 18, 85–92.

Allena, S., Newtson, C.M. 2011. State-of-the-art review on early-age shrinkage of concrete, *Indian Concrete Journal*, 86(7), 14–20.

Alonso, J.L., Wesche, K. 1991. Characterization of fly ash, in *Fly Ash in Concrete Properties and Performance* (Ed. K. Wesche), E.&F.N. Spon, London.

Altan, E., Erdoğan, S.T. 2012. Alkali activation of a slag at ambient and elevated temperatures, *Cement and Concrete Composites*, 34, 131–139.

Andiç-Çakır, Ö. 2007. Investigation of test methods on alkali–aggregate reaction, PhD thesis, School of Natural and Applied Sciences, Ege University, Izmir. (in Turkish).

Angst, U.M., Elsener, B., Larsen, C.K., Vennesland, Ø. 2011. Chloride-induced reinforcement corrosion: Electrochemical monitoring of initiation stage and chloride threshold values, *Corrosion Science*, 53, 1451–1464.

Anon. 1914. An investigation of the pozzolanic nature of coal ashes, *Engineering News*, 71(24), 1334–1335.

Antiohos, S.K., Papadakis, V.G., Chaniotakis, E., Tsimas, S. 2007. Improving the performance of ternary blended cements by mixing different types of fly ashes, *Cement and Concrete Research*, 37, 877–885.

Aponte, D.F., Barra, M., Vàzquez, E. 2012. Durability and cementing efficiency of fly ash concretes, *Construction and Building Materials*, 30, 537–546.

Aprianti, E., Shafigh, P., Bahri, S., Farahani, J.N. 2015. Supplementary cementitious materials origin from agricultural wastes – A review, *Construction and Building Materials*, 74, 176–187.

Ardoğa, M.K. 2014. Effect of particle size on heat of hydration of pozzolan-incorporated cements, MS thesis, Middle East Technical University, Ankara, Turkey, 109pp.

Arjunan, P., Silsbee, M.R., Roy, D.M. 1999. Sulfoaluminate-belite cement from low-calcium fly ash and sulfur-rich and other industrial by-products, *Cement and Concrete Research*, 29, 1305–1311.

Aruntaş, H.Y. 1996. Usability of diatomites as pozzolans in cementitious systems, PhD thesis, Gazi University Institute of Natural and Applied Sciences, Ankara, Turkey (in Turkish).

Aruntaş, H.Y., Albayrak, M., Saka, H.A., Tokyay, M. 1998. Investigation of diatomite properties from Ankara-Kızılcahamam and Çankırı-Çerkeş Regions, *Turkish Journal of Engineering and Environmental Science*, 22, 337–343.

ASTM C 91. 2003. Standard specification for masonry cement. ASTM, 100 Barr Harbor Drive, West Conshohocken, PA.

ASTM C 94/C 94 M-00. 2000. Standard specification for ready-mixed concrete. ASTM, 100 Barr Harbor Drive, West Conshohocken, PA.

ASTM C 109. 2013. Standard test method for compressive strength of hydraulic cement mortars (using 2 in. or [50 mm] Cube Specimens). ASTM, 100 Barr Harbor Drive, West Conshohocken, PA.

ASTM C 125. 2000. Standard terminology relating to concrete and concrete aggregates. ASTM, 100 Barr Harbor Drive, West Conshohocken, PA.

ASTM C 150. 2012. Standard specification for portland cement. ASTM, 100 Barr Harbor Drive, West Conshohocken, PA.

ASTM C 186. 2015. Standard test method for heat of hydration of hydraulic cement. ASTM, 100 Barr Harbor Drive, West Conshohocken, PA.

ASTM C 191. 2008. Standard test methods for time of setting of hydraulic cement by vicat needle. ASTM, 100 Barr Harbor Drive, West Conshohocken, PA.

ASTM C 219. 2001. Standard terminology relating to hydraulic cement. ASTM, 100 Barr Harbor Drive, West Conshohocken, PA.

ASTM C 266. 2013. Standard test methods for time of setting of hydraulic cement paste by gillmore needles. ASTM, 100 Barr Harbor Drive, West Conshohocken, PA.

ASTM C 403. 2008. Standard test method for time of setting of concrete mixtures by penetration resistance. ASTM, 100 Barr Harbor Drive, West Conshohocken, PA.

ASTM C 441. 1996. Standard test method for effectiveness of mineral admixtures or ground blast furnace slag in preventing excessive expansion of concrete due to the alkali–silica reaction. ASTM, 100 Barr Harbor Drive, West Conshohocken, PA.

ASTM C 452. 1985. Standard test method for potential expansion of portland cement mortars exposed to sulfate. ASTM, 100 Barr Harbor Drive, West Conshohocken, PA.

ASTM C 595. 2014. Standard specification for blended hydraulic cements. ASTM, 100 Barr Harbor Drive, West Conshohocken, PA.

ASTM C 618. 2012. Standard specification for coal fly ash and raw or calcined natural pozzolan for use as a mineral admixture in concrete. ASTM, 100 Barr Harbor Drive, West Conshohocken, PA.

ASTM C 642. 2004. Standard test method for density, absorption, and voids in hardened concrete. ASTM, 100 Barr Harbor Drive, West Conshohocken, PA.

ASTM C 672. 1998. Standard test method for scaling resistance of concrete surfaces exposed to deicing chemicals. ASTM, 100 Barr Harbor Drive, West Conshohocken, PA.

ASTM C 989. 2014. Standard specification for slag cement for use in concrete and mortars. ASTM, 100 Barr Harbor Drive, West Conshohocken, PA.

ASTM C 1012. 2013. Standard test method for length change of hydraulic cement mortars exposed to a sulfate solution. ASTM, 100 Barr Harbor Drive, West Conshohocken, PA.

ASTM C 1157. 2011. Standard performance specification for hydraulic cement. ASTM, 100 Barr Harbor Drive, West Conshohocken, PA.

ASTM C 1202. 2011. Standard test method for electrical indication of concrete's ability to resist chloride ion penetration. ASTM, 100 Barr Harbor Drive, West Conshohocken, PA.

ASTM C 1240. 2014. Standard specification for silica fume used in cementitious mixtures. ASTM, 100 Barr Harbor Drive, West Conshohocken, PA.

ASTM C 1260. 1994. Standard test method for potential reactivity of aggregates (Mortar-Bar Method). ASTM, 100 Barr Harbor Drive, West Conshohocken, PA.

ASTM C 1585. 2004. Standard test method for measurement of rate of absorption of water by hydraulic cement concretes. ASTM, 100 Barr Harbor Drive, West Conshohocken, PA.

Atiş, C.D. 2003. Accelerated carbonation and testing of concrete made with fly ash, *Construction and Building Materials*, 17, 147–152.

Atzeni, C., Massidda, L., Sanna, U. 1987. Effect of pore size distribution on strength of hardened cement pastes, *Proceedings of the 1st International RILEM Congress, Pore Structure and Material Properties* (Ed. J.C. Mosa), Chapman-Hall, Versailles, pp. 195–202.

Aubert, J.E., Husson, B., Vaquier, A. 2004. Use of municipal solid waste incineration fly ash in concrete, *Cement and Concrete Research*, 34, 957–963.

Aydın, F., Ardalı, Y. 2012. Seawater desalination technologies, *Sigma Journal of Engineering and Natural Sciences*, 30, 156–178.

Aydın, S., Baradan, B. 2014. Effect of activator type and content on properties of alkali-activated slag mortars, *Composites: Part B*, 57, 166–172.

Babu, K.G., Kumar, V.S.R. 2000. Efficiency of GGBS in concrete, *Cement and Concrete Research*, 30, 1031–1036.

Babu, K.G., Rao, G.S.N. 1996. Efficiency of fly ash in concrete with age, *Cement and Concrete Research*, 26(3), 465–474.

Babu, S.D., Babu, K.G., Wee, T.H. 2005. Properties of lightweight expanded polystyrene aggregate concretes containing fly ash, *Cement and Concrete Research*, 35, 1218–1223.

Bai, J., Wild, S., Sabir, B.B. 2003. Chloride ingress and strength loss in concrete with different PC-PFA-MK binder compositions exposed to synthetic seawater, *Cement and Concrete Research*, 33, 353–362.

Baltrus, J.P., LaCount, R.B. 2001. Measurement of adsorption of air-entraining admixture on fly ash in concrete and cement, *Cement and Concrete Research*, 31, 819–824.

Banfill, P.F.G., Beaupré, D., Chapdelaine, F., de Larrard, F., Domone, P.L., Nachbaur, L., Sedran, T., Wallevik, J.E., Wallevik, O. 2001. Comparison of concrete rheometers: International tests at LCPC (Nantes, France) in October 2000 (Eds. C.F. Ferraris and L.E. Brower), NISTIR 6819, National Institute of Standards and Technology, Washington, USA.

Barberon, F., Baroghel-Bouny, V., Zanni, H., Bresson, B., d'Espinose de la Caillerie, J.-P., Malosse, L., Gan, Z. 2005. Interactions between chloride and cement-paste materials, *Magnetic Resonance Imaging*, 23, 267–272.

Barneyback, R.S., Diamond, S. 1981. Expression and analysis of pore fluids of hardened cement pastes and mortars, *Cement and Concrete Research*, 11, 279–285.

Baroghel-Bouny, V., Mounanga, P., Khelidj, A., Loukili, A., Rafai, N. 2006. Autogeneous deformations of cement pastes: Part II. W/C effects, micro-macro correlations, and threshold values, *Cement and Concrete Research*, 36(1), 123–136.

Bektaş, F., Turanlı, L., Monteiro, P.J.M. 2005. Use of perlite powder to suppress the alkali–silica reaction, *Cement and Concrete Research*, 35(10), 2014–2017.

Bektaş, F., Wang, K. 2012. Performance of ground clay brick in ASR-affected concrete: Effects on expansion, mechanical properties and ASR gel chemistry, *Cement and Concrete Composites*, 34, 273–278.

Bektaş, F., Wang, K., Ceylan, H. 2008. Use of ground clay brick as a pozzolanic material in concrete, *Journal of ASTM International*, 5(10), 1–10.

Benaicha, M., Roguies, X., Jalbaud, O., Burtschell, Y., Alaoui, A.H. 2015. Influence of silica fume and viscosity modifying agent on the mechanical and rheological behavior of self-compacting concrete, *Construction and Building Materials*, 84, 103–110.

Bensted, J. 1999. Thaumasite – background and nature in deterioration of cements, mortars and concretes, *Cement and Concrete Composites*, 21, 117–121.

Bensted, J., Coleman, N. 2003. Cement and concrete – 7000 BC to 1900 AD, *Cement-Wapno-Beton*, 8/70(3), 134–142.

Bentz, D.P., Clifton, J.R., Ferraris, C.F., Garboczi, E.J. 1999. Transport properties and durability of concrete: Literature review and research plan, NISTIR 6395, *Building and Fire Research Laboratory*, National Institute of Standards and Technology, Maryland, USA, 41pp.

Bentz, D.P., Jensen, O.M. 2004. Mitigation strategies for autogeneous shrinkage cracking, *Cement and Concrete Composites*, 26, 677–685.

Beycioğlu, A., Aruntaş, H.Y. 2014. Workability and mechanical properties of self-compacting concretes containing LLFA, GBFS and MC, *Construction and Building Materials*, 73, 626–635.

Bhat, S.T., Lovell, C.W. 1997. Flowable fill using waste foundry sand: A substitute for compacted or stabilized soil, *Proceedings: Testing Soil Mixed With Waste or Recycled Materials (ASTM STP-1275)*, pp. 26–41. ASTM, Conshohocken, PA, USA.

Biczok, I. 1967. *Concrete Corrosion and Concrete Protection*. Chemical Publishing Company, New York, NY.

Boikova, A.I., Grischenko, L.V., Domanski, A.I. 1980. Hydration of C_3A and solid solutions of various compositions, *Proceedings of 7th International Congress on Chemistry of Cement*, V. IV, pp. 460–464, Paris.

Boukendakdji, O., Kadri, E.-H., Kenai, S. 2012. Effects of granulated blast furnace slag and superplasticizer type on the fresh properties and compressive strength of self-compacting concrete, *Cement and Concrete Composites*, 34, 583–590.

Bouzoubaâ, N., Zhang, M.H., Malhotra, V.M. 2001. Mechanical properties and durability of concrete made with high-volume fly ash blended cements using a coarse fly ash, *Cement and Concrete Research*, 31, 1393–1402.

Braun, H., Gebauer, J. 1983. Possibilities and limits of using fly-ash in cement, *Zement-Kalk-Gips*, 36(5), 254–258.

Broomfield, J.P. 1997. *Corrosion of Steel in Concrete*. E.&F.N. Spon, London.

Buck, A.D. 1987. Use of cementitious materials other than portland cement, in concrete durability, *Proceedings Katherine and Bryant Mather International Conference* (Ed. J.M. Scanlon), ACI SP-100, American Concrete Institute, Detroit, USA, pp. 1863–1881.

Buil, M., Palliére, A.M., Roussel, B. 1984. High-strength mortars containing condensed silica fume, *Cement and Concrete Research*, 14(5), 693–704.

Cabrera, J.G. 1996. Deterioration of concrete due to reinforcement steel corrosion, *Cement and Concrete Composites*, 18, 47–59.

Cabrera, J.G., Claisse, P.A., Hunt, D.N. 1995. A statistical analysis of the factors which contribute to the corrosion of steel in portland cement and silica fume concrete, *Construction and Building Materials*, 9(2), 105–113.

Çalışkan, S. 2003. Aggregate/mortar interface: Influence of silica fume at the micro- and macro-level, *Cement and Concrete Composites*, 25, 557–564.

Carette, G.G., Malhotra, V.M. 1983. Early-age strength development of concrete incorporating fly ash and condensed silica fume, *1st International Conference on Use of Fly Ash, Silica Fume, Slag, and Other Mineral By-products in Concrete, ACI SP-79*, Vol. 2, pp. 765–784.

CEB-FIP. 1990. CEB-FIP Model Code. *Comite Euro-International Du Beton*, Thomas Telford Services Ltd., London, 437pp.

Celik, K., Jackson, M.D., Mancio, M., Meral, C., Emwas, A.-H., Mehta, P.K., Monteiro, P.J.M. 2014. High-volume natural volcanic pozzolan and limestone powder as partial replacements for portland cement in self-compacting and sustainable concrete, *Cement and Concrete Composites*, 45, pp. 136–147.

CEMBUREAU. 2013. The Role of Cement in the 2050 Low-Carbon Economy, CEMBUREAU The European Cement Association, Brussels, 64pp.

CEMBUREAU. 2015. Activity Report 2014, CEMBUREAU The European Cement Association, Brussels, 42pp.

Çetin, C. 2013. Early heat evolution of different-sized portland cements incorporating ground granulated blast furnace slag, MS thesis, Middle East Technical University, Ankara, Turkey, 109pp.

Chatterjee, A.K. 1996a. High belite cements – present status and future technological options: Part I, *Cement and Concrete Research*, 26, 1213–1225.

Chatterjee, A.K. 1996b. Future technological options: Part II, *Cement and Concrete Research*, 26, 1227–1237.

Chen, B., Liu, J. 2008. Experimental application of mineral admixtures in lightweight concrete with high-strength and workability, *Construction and Building Materials*, 22, pp. 655–659.

Cheng-yi, H., Feldman, R.F. 1985. Hydration reactions in portland cement-silica fume blends, *Cement and Concrete Research*, 15(4), 585–592.

Chidiac, S.E., Panesar, D.K. 2008. Evolution of mechanical properties of concrete containing ground granulated blast furnace slag and effects on the scaling resistance test at 28 days, *Cement and Concrete Composites*, 30, 63–71.

Chindaprasirt, P., Homwuttiwong, S., Sirivivatnanon, V. 2004. Influence of fly ash fineness on strength, drying shrinkage and sulfate resistance of blended cement mortar, *Cement and Concrete Research*, 34, 1087–1092.

Cocina, E.V., Morales, E.V., Santos, S.F., Savastano Jr., H., Frias, M. 2011. Pozzolanic behavior of bamboo leaf ash: characterization and determination of the kinetic parameters, *Cement and Concrete Composites*, 33, 68–73.

Çolak, A. 2003. Characteristics of pastes from a portland cement containing different amounts of natural pozzolan, *Cement and Concrete Research*, 33(4), 585–593.

Collepardi, M. 2003. A state-of-the-art review on delayed ettringite attack on concrete, *Cement and Concrete Composites*, 25, 401–407.

Collepardi, M., Baldini, G., Pauri, M., Corradi, M. 1978. The effect of pozzolanas on the tricalcium aluminate hydration, *Cement and Concrete Research*, 8(6), 741–751.

Collepardi, M., Marcialis, A., Massidda, L., Sanna, U. 1976. Low-pressure steam curing of compacted lime–pozzolana mixtures, *Cement and Concrete Research*, 6, 497–506.

Collepardi, M., Monosi, S., Moriconi, G., Corradi, M. 1979. Tetracalcium aluminoferrite hydration in the presence of lime and gypsum, *Cement and Concrete Research*, 9(4), 431–437.

Cook, D.J. 1986a. Natural pozzolanas, in *Cement Replacement Materials* (Ed. R.N. Swamy), pp. 1–39, Surrey University Press, London.

Cook, D.J. 1986b. Calcined clay, shale and other soils, in *Cement Replacement Materials* (Ed. R.N. Swamy), pp. 40–72, Surrey University Press, London.

Cook, D.J. 1986c. Rice husk ash, in *Cement Replacement Materials* (Ed. R.N. Swamy), pp. 171–196, Surrey University Press, London.

Cook, J.E. 1982. Research and application of high-strength concrete using class C fly ash, *Concrete International*, 4(7), 72–80.

Corinaldesi, V., Moriconi, G., Naik, T.R. 2010. Characterization of marble powder for its use in mortar and concrete, *Construction and Building Materials*, 24, 113–117.

Costa, U., Massazza, F. 1974. Factors affecting the reaction with lime of italian pozzolanas, *Proceedings of 6th International Congress on Chemistry of Cement*, Supplementary Papers Section III, Stroyizdat, Moscow, pp. 2–18.

Craeye, B., De Scutter, G., Desmet, B., Vantomme, J., Heirman, G., Vandewalle, L., Cizer, Ö. 2010. Effect of mineral filler type on autogenous shrinkage of self compacting concrete, *Cement and Concrete Research*, 40, 908–913.

CSI. 2005. Cement Sustainability Initiative (CSI) Guidelines for the Selection and Use of Fuels and Raw Materials in the Cement Manufacturing Process, Version 1.0, World Bussiness Council for Sustainable Development, Geneva, 35pp.

Dalziel, J.A., Gutteridge, W.A. 1986. The influence of pulverized-fuel ash upon the hydration characteristics and certain physical properties of a portland cement paste, CCA, Technical Report 560, 28pp.

Damtoft, J.S., Lukasik, J., Herfort, D., Sorrentino, D., Gartner, E.M. 2008. Sustainable development and climate change initiatives, *Cement and Concrete Research*, 38, 115–127.

Davies, R.E. 1954. *Pozzolanic Materials – With Special Reference to Their Use in Concrete Pipe*. Technical Memo. American Concrete Pipe Association, Irving, TX.

Davis, R.E., Carlson, R.W., Kelly, J.W., Davis, H.E. 1937. Properties of cements and concretes containing fly ash, *ACI Journal of Proceedings*, 33(5), 577–612.

Davis, R.E., Kelly, J.W., Troxell, G.E., Davis, H.E. 1935. Properties of mortars and concretes containing portland-pozzolan cements, *Proceedings. Journal of American Concrete Institute*, 32, pp. 80–114.

De Aldecoa, R.I., Ortega, F. 2013. International RCC symposium reviews design and construction technology, Conference Report, *Hydropower and Dams*, Issue 3, 1–9.

Deja, J. 2003. Freezing and de-icing salt resistance of blast furnace slag concretes, *Cement and Concrete Composites*, 25, 357–361.

Delibaş, T. 2012. Effects of granulated blast furnace slag, trass and limestone fineness on the properties of blended cements, MS thesis, The Graduate School of Natural and Applied Sciences, Middle East Technical University, Ankara, Turkey, 72pp.

Dellinghausen, L.M., Gastaldini, A.L.G., Vanzin, F.J., Veiga, K.K. 2012. Total shrinkage, oxygen permeability, and chloride ion penetration in concrete made with white portland cement and blast-furnace slag, *Construction and Building Materials*, 37, 652–659.

Demirdağ, S., Uğur, I., Saraç, S. 2008. The effects of cement/fly ash ratios on the volcanic slag aggregate lightweight concrete masonry units, *Construction and Building Materials*, 22, 1730–1735.

Dhir, R.K. 1986. Pulverized-fuel ash, in *Cement Replacement Materials* (Ed. R.N. Swamy), pp. 197–256, Surrey University Press, Guildford, Surrey, UK.

Dhir, R.K., Limbachiya, M.C., McCarthy, M.J., Chaipanich, A. 2007. Evaluation of portland limestone cements for use in concrete construction, *Materials and Structures*, 40, 459–473.

Diamond, S. 1996. Delayed ettringite formation – processes and problems, *Cement and Concrete Composites*, 18, 205–215.

Diamond, S., Thaulow, N. 1974. A sudy of expansion due to alkali–silica reaction as conditioned by the grain size of the reactive aggregate, *Cement and Concrete Research*, 4, 591–607.

Donatello, S., Palomo, A., Fernández-Jiménez, A. 2013. Durability of very high-volume fly ash cement pastes and mortars in aggressive solutions, *Cement and Concrete Composites*, 38, 12–20.

Dotto, J.M.R., de Abreu, A.G., Dal Molin, D.C.C., Müller, I.L. 2004. Influence of silica fume addition on concretes physical properties and on corrosion behaviour of reinforcement bars, *Cement and Concrete Composites*, 26, 31–39.

Dourdounis, E., Stivanakis, V., Angelopoulos, G.N., Chaniotakis, E., Frogoudakis, E., Papanastasiou, D., Papamantellos, D.C. 2004. High-alumina cement production from FeNi-ERF slag, limestone and diasporic bauxite, *Cement and Concrete Research*, 34, 941–947.

Duan, P., Shui, Z., Chen, W., Shen, C. 2013. Effects of metakaolin, silica fume and slag on pore structure, interfacial transition zone and compressive strength of concrete, *Construction and Building Materials*, 44, 1–6.

Duda, W.H. 1977. *Cement Data-Book*, 2nd Ed., Bauverlag GmbH, Weisbaden.

Dunstan, E.R. 1980. A possible method for identifying fly ashes that will improve the sulfate resistance of concretes, *Cement, Concrete, and Aggregates*, 2(1), 20–30.

Dunstan, E.R. 1981. The effect of fly ash on concrete alkali–aggregate reaction, *Cement, Concrete and Aggregates*, 3, 101–104.

Durand, B., Berard, J., Soles, J.A. 1987. Comparison of the effectiveness of mineral admixtures to counteract alkali–aggregate reaction, in *Concrete Aggregate Reactions* (Ed. P.E. Grattan-Bellew), pp. 30–35, Noyes Publications, New Jersey.

Duvallet, T., Rathbone, R.F., Henke, K.R., Jewell, R.B. 2009. Low-energy low-CO_2-emitting cements produced from coal combustion by-products and red mud. In *2009 World of Coal (WOCA) Conference*, Lexington, KY, USA.

Duyou, L., Fournier, B., Grattan-Bellew, P.E. 2006. Evaluation of accelerated test methods for determining alkali–silica reactivity of concrete aggregates, *Cement and Concrete Composites*, 28, 546–554.

Dyer, T.D., Dhir, R.K. 2004. Hydration reactions of cement combinations containing vitrified incinerator fly ash, *Cement and Concrete Research*, 34, 849–856.

Edmeades, R.M., Hewlett, P.C. 1988. Cement admixtures, in *Lea's Chemistry of Cement and Concrete*, 4th Ed. (Ed. P.C. Hewlett), Elsevier, Oxford.

EFNARC. 2002. *Specification and Guidelines for Self-Compacting Concrete*, European Federation for Specialist Construction Chemicals and Concrete Systems, Norfolk, UK.

Eglinton, M. 1988. Resistance of concrete to destructive agencies, in *Lea's Chemistry of Cement and Concrete*, 4th Ed. (Ed. P.C. Hewlett), Elsevier, Oxford.

Ekolu, S.O., Thomas, M.D.A., Hooton, R.D. 2007. Implications of pre-formed microcracking in relation to the theories of DEF mechanism, *Cement and Concrete Research*, 37, 161–165.

Ellerbrock, H.G., Sprung, S., Kuhlman, K. 1985. Influence of Interground Additives on the Properties of Cement, Zement-Kalk-Gips, 38(11), 586–588.

EN 196-1. 2005. *Methods of Testing Cement – Part 1: Determination of Strength.* CEN, Brussels.

EN 196-3. 2002. *Methods of Testing Cement – Part 3: Determination of Setting Time and Soundness.* CEN, Brussels.

EN 196-8. 2011. *Methods of Testing Cement – Part 8: Heat of Hydration – Solution Method.* CEN, Brussels.

EN 196-9. 2010. *Methods of Testing Cement – Part 9: Heat of Hydration – Semi-adiabatic Method.* CEN, Brussels.

EN 197-1. 2012. *Cement-Part 1: Composition, Specification and Conformity Criteria.* CEN, Brussels.

EN 206-1. 2014. *Concrete-Part 1: Specification, Performance, Production and Conformity.* CEN, Brussels.

EN 450-1. 2015. *Fly Ash for Concrete – Part 1: Definition, Specifications, and Conformity Criteria.* CEN, Brussels.

EN 13263-1. 2010. *Silica Fume for Concrete – Part 1: Definitions, Specifications, and Conformity Criteria.* CEN, Brussels.

EN 15167-1. 2006. *Ground Granulated Blast Furnace Slag for Use in Concrete, Mortar, and Grout – Part 1: Definitions, Specifications, and Conformity Criteria.* CEN, Brussels.

Eppers, S. 2010. Assessing the autogenous shrinkage cracking propensity of concrete by means of the restrained ring test, PhD dissertaion, Technical University of Dresden, 178pp.

Erdem, T.K., Kırca, Ö. 2008. Use of binary and ternary blends in high-strength concrete, *Construction and Building Materials*, 22, 1477–1483.

Erdoğan, T.Y. 2007. *Beton*, 2nd Ed., METU Press, Ankara.

Erdoğan, T.Y., Tokyay, M., Ramyar, K. 1992. Investigations on the sulfate resistance of high-lime fly ash incorporating PC-FA mortars, in *4th CANMET/ACI International Conference on Fly Ash, Silica* Fume, Slag and Natural Pozzolans in Concrete, ACI SP-132 (Ed. V.M. Malhotra), American Concrete Institute, Detroit, MI, USA, V.1, 271–280.

Erdoğdu, K. 1996. Effects of pozzolanic additions on grindability and some mechanical properties of pozzolanic cements of different fineness values, MS thesis, Middle East Technical University, Ankara, Turkey, 96pp.

Erdoğdu, K. 2002. Hydration properties of limestone incorporated cementitious systems, PhD thesis, Middle East Technical University, Ankara, Turkey, 115pp.

Erdoğdu, K., Tokyay, M., Türker, P. 1999. Comparison of intergrinding and separate grinding for the production of natural pozzolan and GBFS-incorporated blended cements, *Cement and Concrete Research*, 29(5), 743–746.

Erdoğdu, K., Türker, P. 1998. Effects of fly ash particle size on strength of portland cement fly ash mortars, *Cement and Concrete Research*, 28(9), 1217–1222.

Erdoğdu, Ş., Arslantürk, C., Kurbetci, Ş. 2011. Influence of fly ash and silica fume on the consistency retention and compressive strength of concrete subjected to prolonged agitating, *Construction and Building Materials*, 25, 1277–1281.

Escalante-Garcia, J.I., Gomez, L.Y., Johal, K.K., Mendoza, G., Mancha, H., Méndez, J. 2001. Reactivity of blast-furnace slag in portland cement blends hydrated under different conditions, *Cement and Concrete Research*, 31, 1403–1409.

Escalante-Garcia, J.I., Sharp, J.H. 2004. The chemical composition and microstructure of hydration products in blended cements, *Cement and Concrete Composites*, 26, 967–976.

Ezziane, K., Bougara, A., Kadri, A., Khelafi, H., Kadri, E. 2007. Compressive strength of mortar containing natural pozzolan under various curing temperature, *Cement and Concrete Composites*, 29, 587–593.

Fajun, W., Grutzeck, M.W., Roy, D.M. 1985. The retarding effects of fly ash upon the hydration of cement pastes: The first 24 hours, *Cement and Concrete Research*, 15(1), 174–184.

Federico, L.M., Chidiac, S.E. 2009. Waste glass as a supplementary cementitious material in concrete – critical review of treatment methods, *Cement and Concrete Composites*, 31(8), 606–610.

Feldman, R.F., Sereda, P.J. 1961. Characteristics of sorption and expansion isotherms of reactive limestone aggregate. *Proceedings of ACI*, 58(2), 203–213.

Felekoğlu, B. 2007. Utilisation of high volumes of limestone quarry wastes in concrete industry (Self-compacting Concrete Case), resources, *Conservation and Recycling*, 51, 770–791.

Felekoğlu, B., Türkel, S., Kalyoncu, H. 2009. Optimization of fineness to maximize the strength activity of high-calcium ground fly ash-portland cement composites, *Construction and Building Materials*, 25, 2053–2061.

Fernandez, R., Martirena, F., Scrivener, K.L. 2011. The origin of pozzolanic activity of calcined clay minerals: A comparison between kaolinite, illite, and montmorillonite, *Cement and Concrete Research*, 41, 113–122.

Ferraris, C.F., Obla, K.H., Hill, R. 2001. The influence of mineral admixtures on the rheology of cement paste and concrete, *Cement and Concrete Research*, 31, 245–255.

Ferreira, C., Riberio, A., Ottosen, L. 2003. Possible applications for municipal solid waste fly ash, *Journal of Hazardous Materials*, B96, 201–216.

Fidjestol, P., Dåstøl, M. 2014. The History of Silica Fume in Concrete-from Knovelty to Key Ingredient in High-Performance Concrete, http://www.ibracon.org.br/eventos/50cbc/plenarias/PER_FIDJESTOL.pdf (last visited: May 20, 2015).

Fidjestøl, P. Lewis, R. 1988. Microsilica as an addition, in *Lea's Chemistry of Cement and Concrete* (Ed. P.C. Hewlett), Elsevier, London.

Filipponi, P., Polettini, A., Pomi, R., Sirini, P. 2003. Physical and mechanical properties of cement-based products containing incineration bottom ash, *Waste Management*, 23, 145–156.

Freisleben Hansen, P., Pedersen, E.J. 1977. Maturity computer for controlled curing and hardening of concrete, *Nordisk Betong*, 1, 19–34.

Fukuhara, M., Goto, S., Asaga, K., Daimon, M., Kondo, R. 1981. Mechanisms and kinetics of C_4AF hydration with gypsum, *Cement and Concrete Research*, 11(3), 407–414.

Gabr, M.A., Bowders, J.J. 2000. Controlled low-strength material using fly ash and AMD sludge, *Journal of Hazardous Materials*, 76, 251–263.

Gartner, E. 2004. Industrially interesting approaches to "Low-CO2" cements, *Cement and Concrete Research*, 34, 1489–1498.

Gillot, J.E. 1986. Alkali reactivity problems with emphasis on canadian aggregates, *Engineering Geology*, 23, 29–43.

Glasser, F.P. 1988. The burning of portland cement, in *Lea's Chemistry of Cement and Concrete*, 4th Ed. (Ed. P.C. Hewlett), pp. 195–236, Elsevier, London.

Glasser, F.P. 1992. Chemistry of alkali–aggregate reaction, in *Alkali–Silica Reaction in Concrete* (Ed. R.N. Swamy), pp. 30–53, Van Nostrand Rheinhold, NewYork.

Gomes, S., François, M. 2000. Characterization of mullite in silicoaluminous fly ash by XRD, TEM and 29SiMAS NMR, *Cement and Concrete Research*, 30, 175–181.

Grattan-Bellew, P.E. 1997. A critical review of ultra-accelerated tests for alkali-silica reactivity, *Cement and Concrete Composites*, 19, 403–414.

Grattan-Bellew, P.E., Mitchell, L.D., Margeson, J., Min, D. 2010. Is alkali–carbonate reaction just a variant of alkali–silica reaction ACR=ASR? *Cement and Concrete Research*, 40, 556–562.

Grist, E.R., Paine, K.A., Heath, A., Norman, J., Pinder, H. 2015. The environmental credentials of hydraulic lime-pozzolan concretes, *Journal of Cleaner Production*, 93, 26–37.

Gruyaert, E., Maes, M., De Belie, N. 2013. Performance of BFS concrete: k-value concept versus equivalent performance concept, *Construction and Building Materials*, 47, 441–455.

Gudmundsson, G., Olafsson, H. 1999. Alkali–silica reactions and silica fume. 20 years of experience in iceland, *Cement and Concrete Research*, 29, 1289–1297.

Guerrero, A., Goñi, S., Campillo, I., Moragues, A. 2004. Belite cement clinker from coal fly ash of high Ca content. Optimization of synthesis parameters, *Environmental Science and Technology*, 38, 3209–3213.

Guerrero, A., Goñi, S., Macías, A., Luxán, M.P. 1999. Hydraulic activity and microstructural characterization of new fly ash-belite cements synthesized at different temperatures, *Journal of Materials Research*, 14(6), 2680–2687.

Güneş, A. 2010. Synthesis of alinite cement using soda solid waste, MS thesis, Middle East Technical University, Ankara, Turkey, 52pp.

Güneş, A., Tokyay, M., Yaman, I.Ö., Öztürk, A. 2012. Properties of alinite cement produced by using soda sludge, *Advances in Cement Research*, 24(1), 1–8.

Gür, N., Aktaş, Y., Civaş, A., Öztekin, E. 2010. Utilization of solid waste of soda ash plant as a mineral additive in cement, *Cement and Concrete World*, 15, 57–69.

Gutteridge, W.A., Dalziel, J.A. 1990a. Filler cement: The effect of the secondary component on the hydration of portland cement: Part 1. *A Fine Nonhydraulic Filler, Cement and Concrete Research*, 20(5), 778–782.

Gutteridge, W.A., Dalziel, J.A. 1990b. Filler cement: The effect of the secondary component on the hydration of portland cement: Part 2. *Fine Hydraulic Binders, Cement and Concrete Research*, 20(6), 853–861.

Halse, Y., Pratt, P.L., Dalziel, J.A., Gutteridge, W.A. 1984. Development of microstructure and other properties in fly ash OPC systems, *Cement and Concrete Research*, 14(4), 491–498.

Hanehara, S., Tomosawa, F., Kobayakawa, M., Hwang, K.R. 2001. Effects of water/powder ratio, mixing ratio of fly ash, and curing temperature on pozzolanic reaction of fly ash in cement paste, *Cement and Concrete Research*, 31(1), 31–39.

Haque, M.N. 1996. Strength development and drying shrinkage of high-strength concretes, *Cement and Concrete Composites*, 18, 333–342.

Hassan, A.A.A., Lachemi, M., Hossain, K.M.A. 2012. Effect of metakaolin and silica fume on the durability of self-consolidating concrete, *Cement and Concrete Composites*, 34, 801–807.

Hawkins, P., Tennis, P., Detwiler, R. 2003. *The Use of Limestone in Portland Cement: A State-of-the-Art Review*, PCA, Skokie, IL.

Heidrich, C., Feuerborn, H-J., Weir, A. 2013. Coal combustion products: A global perspective, *2013 World of Coal Ash (WOCA) Conference*, Lexington, Kentucky, USA, 22–25.

Heirman, G., Vandewalle, L., Van Gemert, D., Boel, V., Audenaert, K., De Schutter, G., Desmet, B., Vantomme, J. 2008. Time-dependent deformations of limestone powder type self-compacting concrete, *Engineering Structures*, 30, 2945–2956.

Helmuth, R.A. 1986. Water-reducing properties of fly ash cement pastes, *Mortars, and Concretes: Causes and Test Methods*, ACI SP-91, pp. 723–740.

Hime, W.G., Mather, B. 1999. "Sulfate Attack", or Is It? *Cement and Concrete Research*, 29, 789–791.

Hjorth, J., Skibsted, J., Jakobsen, H.J. 1988. 29Si Mas NMR studies of portland cement components and effects of microsilica on the hydration reaction, *Cement and Concrete Research*, 18, 789–798.

Hobbs, D.W. 1988. *Alkali–Silica Reaction in Concrete*. Thomas Telford, London, 183p.

Hobbs, D.W., Gutteridge, W.A. 1979. Particle size of aggregate and its influence upon the expansion caused by the alkali–silica reaction, *Magazine of Concrete Research*, 31, 235–242.

Holten, C.L.M., Stein, H.N. 1977. Influence of quartz surfaces on the reaction $C_3A+CaSO_4 \cdot 2H_2O+water$, *Cement and Concrete Research*, 7(3), 291–296.

Hooton, R.D. 1993. Influence of silica fume replacement of cement on physical properties and resistance to sulfate attack, freezing and thawing, and alkali–silica reactivity, *ACI Materials Journal*, 90(2), 143–151.

Hossack, A.M., Thomas, M.D.A. 2015. Varying fly ash and slag contents in portland limestone cement mortars exposed to external sulfates, *Construction and Building Materials*, 78, 333–341.

Hossain, K.M.A. 2008. Pumice-based blended cement concretes exposed to marine environment: Effects of mix composition and curing conditions, *Cement and Concrete Composites*, 30, 97–105.

Hubbard, F.H., Dhir, R.K. 1984. A compositional index of the pozzolanic potential of pulverised-fuel ash, *Materials Science Letters*, 3, 958–960.

Iacobescu, R.I., Koumpouri, D., Pontikes, Y., Şaban, R., Angelopoulos, G. 2011. Utilization of EAF metallurgical slag in "Green" belite cement, *U.P.B. Science Bulletin, Series B*, 73(1), 187–194.

Ichikawa, T., Miura, M. 2007. Modified model of alkali–silica reaction, *Cement and Concrete Research*, 37, 1291–1297.

IEA. 2007. *Tracking Industrial Energy Efficiency and CO_2 Emissions*, International Energy Agency, Paris, France, 321pp.

Imbabi, M.S., Carrigan, C., McKenna, S. 2012. Trends and developments in green cement and concrete technology, *International Journal of Sustainable Built Environment*, 1, 194–216.

Inan Sezer, G. 2007. Effects of limestone and clinker properties on the properties of limestone blended cement, PhD thesis, Ege University, Institute of Applied Sciences, Izmir, Turkey (in Turkish).

Irassar, E.F. 2009. Sulfate attack on cementitious materials containing limestone filler – a review, *Cement and Concrete Research*, 39, 241–254.

Isaia, G.C., Gastaldini, A.L.G., Moraes, R. 2003. Physical and pozzolanic action of mineral additions on the mechanical strength of high-performance concrete, *Cement and Concrete Composites*, 25, 69–76.

Itim, A., Ezziane, K., Kadri, E.-H. 2011. Compressive strength and shrinkage of mortar containing various amounts of mineral additions, *Construction and Building Materials*, 25, 3603–3609.

James, J., Rao, M.S. 1986. Characterization of silica in rice husk ash, *American Ceramic Society Bulletin*, 65(8), 1177–1180.

Jani, Y., Hogland, W. 2014. Waste glass in the production of cement and concrete – a review, *Journal of Environmental Chemical Engineering*, 2(3), 1767–1775.

Jastrzebski, D. 1959. *Nature and Properties of Engineering Materials*, John Wiley & Sons Inc., New York.

Jawed, I., Skalny, J., Bach, T., Schubert, P., Bijen, J., Grube, H., Nagataki, S., Ohga, H., Ward, M.A. 1991. Hardened mortar and concrete with fly ash, fly ash in concrete – properties and performance, *Rept. Tech. Comm.* 67-FAB RILEM (Ed. K. Wesche), E.&F.N. Spon, London.

Jensen, O.M., Hansen, P.F. 2001. Autogeneous deformation and RH-change in perspective, *Cement and Concrete Research*, 31, 1869–1865.

Jiang, C., Yang, Y., Wang, Y., Zhou, Y., Ma, C. 2014. Autogenous shrinkage of high-performance concrete containing mineral admixtures under different curing conditions, *Construction and Building Materials*, 61, 260–269.

Jiang, L. 1999. The Interfacial transition zone and bond strength between aggregates and cement pastes incorporating high volumes of fly ash, *Cement and Concrete Composites*, 21, 313–316.

Jitchaiyaphum, K., Sinsiri, T., Chindaprasirt, P. 2011. Cellular lightweight concrete containing pozzolan materials, *Procedia Engineering*, 14, 1157–1164.

Johansson, S., Andersen, P.J. 1990. Pozzolanic activity of calcined moler clay, *Cement and Concrete Research*, 20, 447–452.

Joshi, R.C., Ward, M.A. 1980. Cementitious fly ashes – structural and hydration mechanism, *Proceedings of the 7th International Congress on Chemistry of Cement*, Vol 3, pp. IV/78–IV/82, Paris.

Juenger, M.C.G., Winnefeld, F., Provis, J.L., Ideker, J.H. 2011. Advances in alternative cementitious binders, *Cement and Concrete Research*, 42, 1232–1243.

Kalyoncu, R.S. 2000. Slag-iron and steel, in *U.S. Geological Survey Minerals Yearbook*, USGS, Pittsburg, PA, USA, pp. 71.1–71.3.

Kasai, Y., Tobinai, K., Asakura, E, Feng, N. 1992. Comparative study of natural zeolites and other inorganic admixtures in terms of characterization and properties of mortars, *Proceedings of 4th International Conference on Fly Ash, Silica Fume, Slag, and Natural Pozzolans in Concrete, Istanbul*, ACI Special Publication 132, Vol. 1, pp. 615–634.

Kasap, Ö. 2002. Effects of cement type on concrete maturity, MS thesis, Middle East Technical University, Ankara, Turkey, 96pp.

Katayama, T. 2010. The so-called alkali–carbonate reaction (ACR) – its mineralogical and geochemical details, with special reference to ASR, *Cement and Concrete Research*, 40, 643–675.

Kayali, O., Zhu, B. 2005. Corrosion performance of medium-strength and silica fume high-strength reinforced concrete in a chloride solution, *Cement and Concrete Composites*, 27, 117–124.

Kayapınar, O.Ü. 1991. Use of silicoferrochromium slag and ferrochromium condensed silica fume as partial replacement of cement, MS thesis, Middle East Technical University, Ankara, Turkey, 63pp.

Khan, M.I., Siddique, R. 2011. Utilization of silica fume in concrete: Review of durability properties, *Resources, Conservation and Recycling*, 57, 30–35.

Khatri, R.P., Sirivivatnanon, V., Gross, W. 1995. Effect of different supplementary cementitious materials on mechanical properties of high-performance concrete, *Cement and Concrete Research*, 25(1), 209–220.

Khayat, K.H. 1999. Workability, testing and performance of self-consolidating concrete, ACI *Materials Journal*, 96(3), 346–353.

Kılıç, A., Atiş, C.D., Yaşar, E., Özcan, F. 2003. High-strength lightweight concrete made with scoria aggregate containing mineral admixtures, *Cement and Concrete Research*, 33, 1595–1599.

Kim, J.-K., Kim, J.-S., Ha, G.J., Kim, Y.Y. 2007. Tensile and fiber dispersion performance of ECC (Engineered Cementitious Composites) produced with ground granulated blast furnace slag, *Cement and Concrete Research*, 37, 1096–1105.

Kıyak, B., Özer, A., Altundoğan, H.S., Erdem, M., Tümen, F. 1999. Cr(VI) reduction in aqueous solutions by using copper smelter slag, *Waste Management*, 19, 333–338.

Krstulović, P., Kamenić, N., Popović, K. 1994. A new approach in evaluation of filler effect in cement. I. effect on strength and workability of mortar and concrete, *Cement and Concrete Research*, 24(4), 721–727.

Külaots, I., Hsu, A., Hurt, R.H., Suuberg, E.M. 2003. Adsorption of surfactants on unburned carbon in fly ash and development of a standardized foam index test, *Cement and Concrete Research*, 33, 2091–2099.

Kumar, A., Rao, A.U., Sabhahit, N. 2013. Reactive powder concrete properties with cement replacement using waste materials, *International Journal of Scientific and Engineering Research*, 4(5), 203–206.

Kumar, R., Bhattacharjee, B. 2003. Porosity, pore size distribution and *in situ* strength of concrete, *Cement and Concrete Research*, 33, 155–164.

Kwon, Y-j. 2005. A study on the alkali–aggregate reaction in high-strength concrete with particular respect to the ground granulated blast-furnace slag effect, *Cement and Concrete Research*, 35, 1305–1313.

LaBarca, I.K., Foley, R.D., Cramer, S.M. 2007. Effects of ground granulated blast furnace slag in portland cement concrete (PCC)-expanded study, SPR # 0092-05-01, Wisconsin Highway Research Program, The Wisconsin Department of Transportation, 77pp.

Lachemi, M., Hossain, K.M.A., Shehata, M., Thaha, W. 2008. Controlled low-strength materials incorporating cement kiln dust from various sources, *Cement and Concrete Composites*, 30, 381–391.

Lachemi, M., Şahmaran, M., Hossain, K.M.A., Lofty, A., Shehata, M. 2010. Properties of controlled low-strength materials incorporating cement kiln dust and slag, *Cement and Concrete Composites*, 32, 623–629.

Laibao, L., Yunsheng, Z., Wenhua, Z., Zhiyong, L., Lihua, Z. 2013. Investigating the influence of basalt as a mineral admixture on hydration and microstructure formation mechanism of cement, *Construction and Building Materials*, 48, 434–440.

Lam, L., Wong, Y.L., Poon, C.S. 2000. Degree of hydration and gel/space ratio of high volume fly ash/cement systems, *Cement and Concrete Research*, 30, 747–756.

Lange, F., Mörtel, N., Rudert, V. 1997. Dense packing of cement pastes and resulting consequences on mortar properties, *Cement and Concrete Research*, 27(10), 1481–1488.

Laskar, A.I., Talukdar, S. 2008. Rheological behavior of high-performance concrete with mineral admixtures and their blending, *Construction and Building Materials*, 22, 2345–2354.

Lawrence, C.D. 1988. The production of low-energy cements, in *Lea's Chemistry of Cement and Concrete*, 4th Ed. (Ed. P.C. Hewlett), 421–470, Elsevier, Oxford.

Lawrence, P., Cyr, M., Ringot, E. 2003. Mineral admixtures in mortars: Effect of inert materials on short-term hydration, *Cement and Concrete Research*, 33(12), 1939–1947.

Lawrence, P., Cyr, M., Ringot, E. 2005. Mineral admixtures in mortars effect of type, amount and fineness of fine constituents on compressive strength, *Cement and Concrete Research*, 35, 1092–1105.

Lewis, R., Sear, L., Wainwright, P., Ryle, R. 2003. Cementitious additions, in *Advanced Concrete Technology Set*, V.3, Butterworth-Heinemann, London.

Lewis, S. 2015. The World's Largest Roller-Compacted Concrete Dams, http://enr.construction.com/infrastructure/water_dams/2015/0331-Roller-Compacted-Concrete-Dams.asp, (last visited: June 3, 2015).

Li, H., Wee, T.H., Wong, S.F. 2002. Early-age creep and shrinkage of blended cement concrete, *ACI Materials Journal*, 99(1), 3–10.

Li, J. Yao, Y. 2001. A study on creep and drying shrinkage of high-performance concrete, *Cement and Concrete Research*, 31, 1203–1206.

Li, V.C. 2002. Large volume, high-performance applications of fibers in civil engineering, *Journal of Applied Polymer Science*, 83, 660–686.

Li, Y., Bao, J., Guo, Y. 2010. The relationship between autogenous shrinkage and pore structure of cement paste with mineral admixtures, *Construction and Building Materials*, 24, 1855–1860.

Lian, C., Zhuge, Y., Beecham, S. 2011. The relationship between porosity and strength for porous concrete, *Construction and Building Materials*, 25, 4294–4298.

Litvan, G.G. 1976. Frost action in cement in the presence of de-icers, *Cement and Concrete Research*, 6(3), 351–356.

Liu, H., Lu, Z., Lin, S. 1980. Composition and hydration of high-calcium fly ash, *Proceedings of 7th International Congress on Chemistry of Cement*, Paris, V. III, pp. IV/7-IV/12.

Locher, F.W. 1966. The problem of sulfate resistance of slag cements, *Zement-Kalk-Gips*, 19(9), 395–401.

Lollini, F., Redaelli, E., Bertolini, L. 2014. Effects of portland cement replacement with limestone on the properties of hardened concrete, *Cement and Concrete Composites*, 46, 32–40.

López-Buendía, A.M., Climent, V., Verdú, P. 2006. Lithological influence of aggregate in the alkali–carbonate reaction, *Cement and Concrete Research*, 36, 1490–1500.

Lothenbach, B., Scrivener, K., Hooton, R.D. 2011. Supplementary cementitious materials, *Cement and Concrete Research*, 41(3), 311–323.

Lu, C., Yang, H., Mei, G. 2015. Relationship between slump flow and rheological properties of self compacting concrete with silica fume and its permeability, *Construction and Building Materials*, 75, 157–162.

Ludwig, U., Schwiete, H.E. 1963. Lime combination and new formations in the trass-lime reactions, *Zement-Kalk-Gips*, 6(10), 421–431.

Luke, K., Lachowski, E. 2008. Internal composition of 20-year-old fly ash and slag-blended ordinary portland cement pastes, *Journal of the American Ceramic Society*, 91(12), 4084–4092.

Luping, T. 1986. A study of the quantitative relationship between strength and pore size distribution of porous materials, *Cement and Concrete Research*, 16(1), 87–96.

Ma, B., Li, X., Mao, Y., Shen, X. 2013. Synthesis and characterization of high belite sulfoaluminate cement through rich alumina fly ash and desulfurization gypsum, *Ceramics-Silikáty*, 57(1), 7–13.

Maes, M., De Belie, N. 2014. Resistance of concrete and mortar against combined attack of chloride and sodium sulfate, *Cement and Concrete Composites*, 53, 59–72.

Mailvaganam, N.P., Rixom, M.R. 1999. *Chemical Admixtures for Concrete*, E.&F.N. Spon, London.

Maltais, Y., Marchand, J. 1997. Influence of curing temperature on cement hydration and mechanical strength development of fly ash mortars, *Cement and Concrete Research*, 27(7), 1009–1020.

Malvar, L.J., Lenke, L.R. 2006. Efficiency of fly ash in mitigating alkali–silica reaction based on chemical composition, *ACI Materials Journal*, 103(5), 319–326.

Marjanović, N., Komljenović, M., Baščarević, Z., Nikolić, V., Petrović, R. 2015. Physical-mechanical and microstructural properties of alkali-activated fly ash-blast furnace slag blends, *Ceramics International*, 41, 1421–1435.

Marroccoli, M., Pace, M.L., Telesca, A., Valenti, G.L. 2010. Synthesis of calcium sulfoaluminate cements from Al2O3-rich by-products from aluminum manufacture. In *Proceedings of 2nd International Conference on Sustainable Construction Materials and Technologies* (Eds. J. Zachar, P. Claisse, T.R. Naik, E. Ganjian) Coventry Un. and UWM, Ancona, Main Volume 1, pp. 615–624.

Marwan, T., Péra, J., Ambroise, J. 1992. The action of some aggressive solutions on portland and calcined laterite blended cement concretes. in *4th CANMET/ACI International Conference on Fly Ash, Silica Fume, Slag and Natural Pozzolans in Concrete* (Ed. V.M. Malhotra), ACI SP-132, American Concrete Institute, Detroit, MI, USA, Vol. 1, pp. 763–779.

Massazza, F. 1974. Chemistry of pozzolanic additions and mixed cements. In *Proceedings of 6th International Congress on the Chemistry of Cements*, Sroyizdat, Moscow.

Massazza, F. 1988. Pozzolana and pozzolanic cements. in *Lea's Chemistry of Cement and Concrete*, 4th Ed. (Ed. P.C. Hewlett), Elsevier, Oxford.

Matos, A.M., Sousa-Coutinho, J. 2012. Durability of mortar using waste glass powder as cement replacement, *Construction and Building Materials*, 36, 205–215.

Matsushita, F., Aono, Y., Shibata, S. 2004. Calcium silicate structure and carbonation shrinkage of a tobermorite-based material, *Cement and Concrete Research*, 34, 1251–1257.

Mays, G. 1992. *Durability of Concrete Structures*, E.&F.N. Spon, London.

Mazloom, M., Ramezanianpour, A.A., Brooks, J.J. 2004. Effect of silica fume on mechanical properties of high-strength concrete, *Cement and Concrete Composites*, 26, 347–357.

Meddah, M.S., Lmbachiya, M. C., Dhir, R.K. 2014. Potential use of binary and composite limestone cements in concrete production, *Construction and Building Materials*, 58, 193–205.

Megat Johari, M.A., Brooks, J.J., Kabir, S., Rivard, P. 2011. Influence of supplementary cementitious materials on engineering properties of high-strength concrete, *Construction and Building Materials*, 25, 2639–2648.

Mehta, P.K. 1981. Studies on blended portland cements containing santorin earth, *Cement and Concrete Research*, 11, 507–518.

Mehta, P.K. 1983. *Pozzolanic and Cementitious By-Products as Mineral Admixtures for Concrete – A Critical Review*, in ACI SP-79, American Concrete Institute, Detroit, MI, USA, pp. 1–46.

Mehta, P.K. 1986a. Condensed silica fume. In *Cement Replacement Materials* (Ed. R.N. Swamy), Surrey University Press, Guildford, Surrey, UK, pp. 134–170.

Mehta, P.K. 1986b. Effect of fly ash composition on sulfate resistance of cement, *ACI Journal Proceedings*, 83(6), 994–1000.

Mehta, P.K., Monteiro, P.J.M. 2006. *Concrete*, 3rd Ed., McGraw-Hill, New York.

Meland, I. 1983. Influence of condensed silica fume and fly ash on the heat evolution in cement pastes, *Proceedings CANMET/ACI 1st International Conference on the Use of Fly Ash, Silica Fume, Slag, and Other Mineral By-Products in Concrete*, Vol. II, pp. 656–676, Montebello.

Meral, Ç. 2004. Use of perlite as a pozzolanic addition in blended cement production, MS thesis, Middle East Technical University, Ankara, Turkey, 106pp.

Merriaux, K. Lecomte, A., Degeimbre, R., Darimont, A. 2003. Alkali–silica reactivity with pessimum content on devonian aggregates from Belgian arden massive, *Magazine of Concrete Research*, 55(5), 429–437.

Meusel, J.W., Rose, J.H. 1983. Production of granulated blast furnace slag at sparrows point and the workability and strength potential of concrete incorporating the slag, *1st International Conference on Fly Ash, Silica Fume and Slag in Concrete*, ACI Special. Publication SP-79, Vol. 2, pp. 867–890, Montebello.

Mielenz, R.C., Witte, L.P., Glantz, O.J. 1950. Effect of calcination on natural pozzolans, *ASTM Special Technical Publication*, 99, 43–92.

Miller, F.M., Conway, T. 2003. Use of ground granulated blast furnace slag for reduction of expansion due to delayed ettringite formation, *Cement, Concrete and Aggregates*, 25(2), 59–68.

Mindess, S., Young, J.F. 1981. *Concrete*, Prentice-Hall, Englewood Cliffs, NJ.

Monteiro, P.J.M., Wang, K., Sposito, G., dos Santos, M.C., de Andrade, W.P. 1997. Influence of mineral admixtures on the alkali–aggregate reaction, *Cement and Concrete Research*, 27(12), 1899–1909.

Montemor, M.F., Cunha, M.P., Ferreira, M.G., Simões, A.M. 2002. Corrosion behaviour of rebars in fly ash mortar exposed to carbon dioxide and chlorides, *Cement and Concrete Composites*, 24, 45–53.

Montemor, M.F., Simões, A.M.P., Salta, M.M. 2000. Effect of fly ash on concrete reinforcement corrosion studied by EIS, *Cement and Concrete Composites*, 22, 175–185.

Moosberg-Bustnes, H. 2003. Fine particulate by-products from mineral and metallurgical industries as filler in cement-based materials, PhD thesis, Luleå University of Technology, Department of Chemical and Metallurgical Engineering, Sweden.

Moranville-Regourd, M. 1988. Cements made from blastfurnace slag, in *Lea's Chemistry of Cement and Concrete*, 4th Ed. (Ed. P.C. Hewlett), Elsevier, Oxford.

Nagi, M.A., Okamoto, P.A., Kozikowski, R.L., Hover, K. 2007. Evaluating air-entraining admixtures for highway concrete, NCHRP Report 578, TRB, 50pp.

Naik, T.R., Singh, S.S., Hossain, M.M. 1995. Properties of high-performance concrete systems incorporating large amounts of high-lime fly ash, *Construction and Building Materials*, 9(4), 195–204.

Naiqian, F., Hongwei, J., Enyi, C. 1998. Study on the suppression effect of natural zeolite on expansion of concrete due to alkali–aggregate reaction, *Magazine of Concrete Research*, 50(1), 17–24.

Nair, D.G., Fraaij, A., Klaassen, A.A.K., Kentgens, A.P.M. 2008. A structural investigation relating to the pozzolanic activity of rice husk ashes, *Cement and Concrete Research*, 38, 861–869.

Nanthagopalan, P., Haist, M., Santhanam, M., Müller, H.S. 2008. Investigation on the influence of granular packing on the flow properties of cementitious suspensions, *Cement and Concrete Composites*, 30, 763–768.

Nassif, H.H., Najm, H., Suksawang, N. 2005. Effect of pozzolanic materials and curing methods on the elastic modulus of HPC, *Cement and Concrete Composites*, 27, 661–670.

NEA-OECD, Nuclear Energy Agency. 1979. Exposure to radiation from natural radioactivity in building materials, Report by NEA Group of Experts, OECD, Paris.

Negro, A., Stafferi, L. 1979. The hydration of calciumferrites and calcium aluminoferrites, *Zement-Kalk-Gips*, 32(2), 83–88.

Nehdi, M., Duquette, J., Damatty, A.E. 2003. Performance of rice husk ash produced using a new technology as a mineral admixture in concrete, *Cement and Concrete Research*, 33(8), 1203–1210.

Nehdi, M., Mindess, S., Aïtcin, P.-C. 1998. Rheology of high-performance concrete: Effect of ultrafine particles, *Cement and Concrete Research*, 28(5), 687–697.

Neville, A. 2004. The confused world of sulfate attack on concrete, *Cement and Concrete Research*, 34(8), 1275–1296.

Neville, A.M., Dilger, W.H, Brooks, J.J. 1983. *Creep of Plain and Structural Concrete*, Construction Press, London.

Nguyen, V.-H., Leklou, N., Aubert, J.-E., Mounanga, P. 2013. The effect of natural pozzolans on delayed ettringite formation of the heat-cured mortars, *Construction and Building Materials*, 48, 479–484.

Nili, M., Afroughsabet, V. 2010. Combined effect of silica fume and steel fibers on the impact resistance and mechanical properties of concrete, *International Journal of Impact Engineering*, 37, 879–886.

Nixon, P.J., Gaze, M.E. 1983. The effectiveness of fly ashes and granulated blast furnace slags in preventing AAR,61–68. In *Proceedings of 6th International Conference on Alkalies in Concrete* (Eds. G.M. Idorn and S. Rostam), Copenhagen.

Nixon, P.J., Page, C.L., Bollinghaus, R., Canham, I. 1986. The effect of PFA with a high-alkali content on pore solution composition and alkali–silica reaction, *Magazine of Concrete Research*, 38, 30–35.

Nochaiya, T., Wongkeo, W., Chaipanich, A. 2010. Utilization of fly ash with silica fume and properties of portland cement-fly ash-silica fume concrete, *Fuel*, 89, 768–774.

Nowak-Michta, A. 2013. Water-binder ratio influence on de-icing salt scaling of fly ash concretes, *Procedia Engineering*, 57, 823–829.

Nurse, R.W. 1949. Steam curing of concrete, *Magazine of Concrete Research*, 1(2), 79–88.

Odler, I. 1988. Hydration, setting and hardening of portland cement, in *Lea's Chemistry of Cement and Concrete*, 4th Ed. (Ed. P.C. Hewlett), pp. 241–289, Elsevier, London.

Odler, I. 2000. Cements containing ground granulated blast furnace slag, in *Special Inorganic Cements*, E.&F.N. Spon, London.

Odler, I., Chen, Y. 1996. On the delayed expansion of heat cured portland cement pastes and concretes, *Cement and Concrete Composites*, 18(3), 181–185.

O'Farrel, M., Wild, S., Sabir, B.B. 1999. Resistance to chemical attack of ground brick-PC mortar: Part 1. Sodium sulfate solution, *Cement and Concrete Research*, 29(11), 1781–1790.

Ogawa, K., Uchikawa, H., Takemote, K., Yasui, I. 1980. The mechanism of the hydration in the system C3S-pozzolana, *Cement and Concrete Research*, 10(5), 683–696.

Older, I, Rössler, M. 1985. Investigations on the relationship between porosity, structure, and strength of hydrated portland cement pastes: II: Effects of pore structure and degree of hydration, *Cement and Concrete Research*, 15(4), 401–410.

Öner, A., Akyüz, S. 2007. An experimental study on optimum usage of GGBS for the compressive strength of concrete, *Cement and Concrete Composites*, 29, 505–514.

Öner, A., Akyüz, S., Yıldız, R. 2005. An experimental study on strength development of concrete containing fly ash and optimum usage of fly ash in concrete, *Cement and Concrete Research*, 35, 1165–1171.

Opoczky, L. 1993a, Problems relating to grinding technology and quality when grinding composite cements, *Zement-Kalk-Gips*, 46(3), 136–140.

Opoczky, L. 1993b, Progress of particle size distribution during the intergrinding of a clinker-limestone mixture, *Zement-Kalk-Gips*, 46(12), 648–651.

Osbaeck, B., Jons, E.S. 1980. The influence of the content and distribution of Al_2O_3 on the hydration properties of portland cement, *Proceedings of 7th International Congress on Chemistry of Cement*, Vol. IV, 514–519, Paris.

Osborne, G.J. 1999. Durability of portland blast-furnace slag cement concrete, *Cement and Concrete Composites*, 21, 11–21.

Oueslati, O., Duchesne, J. 2012. The effect of SCMs on the corrosion of rebar embedded in mortars subjected to an acetic acid attack, *Cement and Concrete Research*, 42, 467–475.

Över, D. 2012. Early heat evolution in natural pozzolan-incorporated cement hydration, MS thesis, Middle East Technical University, Ankara, Turkey, 83pp.

Özcan, S. 2008. Bonding efficiency of roller compacted concrete with different bedding mixtures. MS thesis, Middle East Technical University, Ankara, Turkey, 96pp.

Pacheco-Torgal, F., Castro-Gomes, J., Jalali, S. 2008. Alkali-activated Binders: A review Part 1: Historical background, terminology, reaction mechanisms and hydration products, *Construction and Building Materials*, 22, 1305–1314.

Paillére, A.M. (Ed.). 1995. *Application of Admixtures in Concrete*, E.&F.N. Spon, London.

Paillére, A.M., Buil, M., Serrano, J.J. 1989. Effect of fiber addition on the autogeneous shrinkage of silica fume concrete, *ACI Materials Journal*, 86(2), 139–144.

Pal, S.C., Mukherjee, A., Pathak, S.R. 2003. Investigation of hydraulic activity of ground granulated blast furnace slag in concrete, *Cement and Concrete Research*, 33, 1481–1486.

Pane, I., Hansen, W. 2005. Investigation of blended cement hydration by isothermal calorimetry and thermal analysis, *Cement and Concrete Research*, 35, 1155–1164.

Papadakis, V.G. 2000. Effect of supplementary cementing materials on concrete resistance against carbonation and chloride ingress, *Cement and Concrete Research*, 30, 291–299.

Papadakis, V.G., Antiohos, S., Tsimas, S. 2002. Supplementary cementing materials in concrete Part II: A fundamental estimation of efficiency factor, *Cement and Concrete Research*, 32, 1533–1538.

Papadakis, V.G., Tsimas, S. 2002. Supplementary cementing materials in concrete Part I: Efficiency and design, *Cement and Concrete Research*, 32, 1525–1532.

Papayianni, J. 1992. Performance of a high-calcium fly ash in roller compacted concrete, *Proceedings of 4th International Conference on Fly Ash, Silica Fume, Slag and Natural Pozzolans in Concrete*, Istanbul, Vol. 1, pp. 367–386.

Parande, A.K., Chitradevi, R.H., Thangavel, K., Karthikeyan, M.S., Ganesh, B., Palaniswamy, N. 2009. Metakaolin: A versatile material to enhance the durability of concrete – an overview, *Structural Concrete*, 10(3), 125–138.

PCA. 1998. Control of air content in concrete, *Concrete Technology Today*, 19(1), 1–3.

Pedersen, K.H., Jensen, A.D., Skjøth-Rasmussen, M.S., Dam-Johansen, K. 2008. A review of the interference of carbon containing fly ash with air entrainment in concrete, *Progress in Energy and Combustion Science*, 34, 135–151.

Péra, J., Momtazi, A.S. 1992. Pozzolanic activity of calcined red mud, in *4th CANMET/ACI International Conference on Fly Ash, Silica Fume, Slag and Natural Pozzolans in Concrete* (Ed. V.M. Malhotra), ACI SP-132, Vol. 1, pp. 749–761.

Pereira-de-Oliveira, L.A., Castro-Gomes, J.P., Santos, P.M.S. 2012. The potential pozzolanic activity of glass and red-clay ceramic waste as cement mortar components, *Construction and Building Materials*, 31, 197–203.

Peter, M.A., Muntean, A., Meier, S.A., Böhm, M. 2008. Competition of several carbonation reactions in concrete: A parametric study, *Cement and Concrete Research*, 38, 1385–1393.

Pietersen, H.S., Kentgens, A.P.M., Nachtegaal, G.H., Veeman, W.S., Bijen, J.M. 1992. Wet-shotcrete for refractory castables, *Proceedings of 4th International Conference on Fly Ash, Silica Fume, Slag and Natural Pozzolans in Concrete*, Istanbul, pp. 795–812.

Pigeon, M., Talbot, C., Marchand, J., Hornain, H. 1996. Surface microstructure and scaling resistance of concrete, *Cement and Concrete Research*, 10, 1555–1566.

Pimraksa, K., Hanjitsuwan, S., Chindaprasirt, P. 2009. Synthesis of belite cement from lignite fly ash, *Ceramics International*, 35, 2415–2425.

Pistilli, M.F., Wintersteen, R., Cechner, R. 1984. The uniformity and influence of silica fume source on the properties of portland cement concrete, cement, *Concrete and Aggregate*, 6(2), 120–124.

Placet, M., Fowler, K. 2002. Towards a sustainable cement industry. *Substudy 7: How Innovation Can Help the Cement Industry Move Toward More Sustainable Practices*, An Independent Study Commissioned by WBCSD, 40pp.

Plowman, C., Cabrera, J.G. 1984. Mechanism and kinetics of hydration of C_3A and C_4AF extracted from cement, *Cement and Concrete Research*, 14(2), 238–248.

Pommersheim, J., Chang, J. 1986. Kinetics of hydration of tricalcium aluminate, *Cement and Concrete Research*, 16(3), 440–450.

Poole, A. B. 1992. Introduction to alkali–aggregate reaction in concrete, in *The Alkali–Silica Reaction in Concrete*, (Ed. R. N. Swamy), 30–53, Van Nostrand Reinhold, New York.

Poon, C.S., Kou, S.C, Lam, K. 2006. Compressive strength, chloride diffusivity and pore structure of high-performance metakaolin and silica fume concrete, *Construction and Building Materials*, 20, 858–865.

Poon, C.S., Lam, L., Wong, Y.L. 2000. A study on high-strength concrete prepared with large volumes of low-calcium fly ash, *Cement and Concrete Research*, 30, 447–455.

Powers, T.C. 1958. Structure and physical properties of hardened portland cement paste, *Journal of American Ceramic Society*, 41(1), 1–6.

Pradip, A., Kapur, P.C. 2004. Manufacture of eco-friendly and energy-efficient alinite cements from fly ashes and other bulk wastes, *Resources Processing*, 51(1), 8–13.

Ramachandran, V.S. 1995. *Concrete Admixtures Handbook*, Noyes Publications, NJ, USA.

Ramezanianpour, A.A., Ghiasvand, E., Nickseresht, I., Mahdikhani, M., Moodi, F. 2009. Influence of various amounts of limestone powder on performance of portland limestone cement concretes, *Cement and Concrete Composition*, 31, 715–720.

Ramlochan, T., Thomas, M., Gruber, K.A. 2000. The effect of metakaolin on alkali–silica reaction in concrete, *Cement and Concrete Research*, 30, 339–344.

Ramlochan, T., Thomas, M.D.A., Hooton, R.D. 2003. The effect of pozzolans and slag on the expansion of mortars cured at elevated temperature, Part I: Expansive behavior, *Cement and Concrete Research*, 33(6), 807–814.

Ramlochan, T., Thomas, M.D.A., Hooton, R.D. 2004. The effect of pozzolans and slag on the expansion of mortars cured at elevated temperature, Part II: Microstructural and microchemical investigations, *Cement and Concrete Research*, 34(8), 1341–1356.

Ramyar, K. 1993. Effects of turkish fly ashes on the portland cement-fly ash systems, PhD thesis, Middle East Technical University, Ankara, Turkey, 208pp.

Ramyar, K., Topal, A., Andiç, Ö. 2005. Effects of aggregate size and angularity on alkali–silica reaction, *Cement and Concrete Research*, 35(11), 2165–2169.

Rao, G.A. 1998. Influence of silica fume replacement of cement on expansion and drying shrinkage, *Cement and Concrete Research*, 28(10), 1505–1509.

Rao, G.A. 2003. Investigations on the performance of silica fume-incorporated cement pastes and mortars, *Cement and Concrete Research*, 33(11), 1765–1770.

Rashad, A.M. 2013. Metakaolin as cementitious material: History, scours, production and composition – A comprehensive overview, *Construction and Building Materials*, 41, 303–318.

Regourd, M. 1986. Slags and slag cements, in *Cement Replacement Materials*, (Ed. R.N. Swamy), Surrey University Press, Guildford, Surrey, UK, pp. 73–99.

Regourd, M., Thomassin, J.H., Baillif, P., Touray, J.C. 1983 Blast furnace slag hydration. *Surface Analysis, Cement and Concrete Research*, 13(4), 549–556.

Rémond, S., Pimienta, P., Bentz, D.P. 2002. Effects of the incorporation of municipal solid waste incineration fly ash in cement pastes and mortars I. experimental study, *Cement and Concrete Research*, 32, 303–311.

Richard, P., Cheyrezy, M. 1995. Composition of reactive powder concretes, *Cement and Concrete Research*, 25(7), 1501–1511.

Richardson, I.G. 1999. The nature of C-S-H in hardened cements, *Cement and Concrete Reserach*, 29(8), 1131–1147.

Richardson, I.G., Groves, G.W. 1997. The structure of calcium silicate hydrate phases present in hardened pastes of white portland cement blast furnace slag blends, *Journal of Materials Science*, 32(18), 4793–4802.

Rodriguez-Camacho, R.E., Uribe-Afif, R. 2002. Importance of using the natural pozzolans on concrete durability, *Cement and Concrete Research*, 32, 1851–1858.

Rønning, T.F. 2001. Freeze-thaw resistance of concrete effect of: curing conditions, moisture exchange and materials, PhD thesis, The Norwegian Institute of Technology, Trondheim.

Saad, M.N.A., de Andrade, W.P., Paulon, V.A. 1982. Properties of mass concrete containing an active pozzolan made from clay, *Concrete International*, 4(7), 59–65.

Sabir, B.B., Wild, S., Bai, J. 2001. Metakaolin and calcined clays as pozzolans for concrete: A review, *Cement and Concrete Composites*, 23, 441–454.

Šahinagić-Isović, M., Markovski, G., Ćećez, M. 2012. Shrinkage strains of concrete – causes and types, *Građevinar*, 64(9), 727–734.

Şahmaran, M., Christianto, H.A., Yaman, İÖ. 2006. The effect of chemical admixtures and mineral additives on the properties of self-compacting mortars, *Cement and Concrete Composites*, 28, 432–440.

Şahmaran, M., Erdem, T.K., Yaman, I.O. 2007. Sulfate resistance of plain and blended cements exposed to wetting-drying and heating-cooling environments, *Construction and Building Materials*, 21(8), 1771–1778.

Şahmaran, M., Lachemi, M., Hossain, K.M.A., Li, V.C. 2009. Internal curing of engineered cementitious composites for prevention of early age autogeneous shrinkage cracking, *Cement and Concrete Research*, 39, 893–901.

Şahmaran, M., Li, V.C. 2008. Durability of mechanically loaded engineered cementitious composites under highly alkaline environments, *Cement and Concrete Composites*, 30, 72–81.

Şahmaran, M., Yıldırım, G., Erdem, T.K. 2013. Self-healing capability of cementitious composites incorporating different supplementary cementitious materials, *Cement and Concrete Composites*, 35, 89–101.

Saikia, N., Cornelis, G., Mertens, G., Elsen, J., Van Balen, K., Van Gerven, T., Vandecasteele, C. 2008. Assessment of Pb-slag, MSWI bottom ash and boiler and fly ash for using as a fine aggregate in cement mortar, *Journal of Hazardous Materials*, 154, 766–777.

Sakai, E., Hoshimo, S., Ohba, Y., Daimon, M. 1997. The fluidity of cement paste with various types of inorganic powders, *Proceedings of 10th International Congress on Chemistry of Cement*, pp. 2ii002–010, Stockholm.

Sakai, E., Miyahara, S., Ohsawa, S., Lee, S.H., Daimon, M. 2005. Hydration of fly ash cement, *Cement and Concrete Research*, 35, 1135–1140.

Samarin, A., Munn, R.L., Ashby, J.B. 1983. The use of fly ash in concrete – Australian experience, *Fly Ash, Silica Fume, Slag and Other Mineral By-Products in Concrete*, ACI SP-79, 143–172.

Santhanam, M., Cohen, M.D., Olek, J. 2001. Sulfate attack research – whither now? *Cement and Concrete Research*, 31, 845–851.

Sata, V., Jaturapitakkul, C., Kiattikomol, K., Influence of pozzolan from various by-product materials on mechanical properties of high-strength concrete, *Construction and Building Materials*, 21, 1589–1598.

Saul, A.G.A. 1951. Principles underlying the steam curing of concrete at atmospheric pressure, *Magazine of Concrete Research*, 2(6), 127–140.

Scali, M.J., Chin, D., Berke, N.S. 1987. Effect of microsilica and fly ash upon the microstructure and permeability of concrete, *Proceedings of 9th International Conference on Cement Microscopy*, Reno, pp. 375–387.

Schmidt, M. 1992. Cement with interground additives, *Zement-Kalk-Gips*, 45(2), 87–92.

Scrivener, K.L., Nonat, A. 2011. Hydration of cementitious materials, present and future, *Cement and Concrete Research*, 41(7), 651–665.

Sersale, R., Rebuffat, P. 1970. Microscopic investigations of hardened lime-pozzolana pastes, *Zement-Kalk-Gips*, 23(4), 182–184.

Sfikas, I.P., Badogiannis, E.G., Trezos, K.G. 2014. Rheology and mechanical characteristics of self-compacting concrete mixtures containing metakaolin, *Construction and Building Materials*, 64, 121–129.

Shafiq, N., Nuruddin, M.F., Khan, S.U., Ayub, T. 2015. Calcined kaolin as cement replacement material and its use in high-strength concrete, *Construction and Building Materials*, 81, 313–323.

Shannag, M.J. 2000. High-strength concrete containing natural pozzolan and silica fume, *Cement and Concrete Composites*, 22, 399–406.

Shannag, M.J. 2011. Characteristics of lightweight concrete containing mineral admixtures, *Construction and Building Materials*, 25, 658–662.

Shannag, M.J., Shaia, H.A. 2003. Sulfate resistance of high-performance concrete, *Cement and Concrete Composites*, 25, 363–369.

Shannag, M.J., Yeginobali, A. 1995. Properties of pastes, mortars and concretes containing natural pozzolan, *Cement and Concrete Research*, 25(3), 647–657.

Shatov, A.A., Dryamina, M.A., Badertdinov, R.N. 2004. Potential utilization of soda production wastes, *Chemistry for Sustainable Development*, 12, 565–571.

Sheen, Y-N., Zhang, L-H., Le, D-H. 2013. Engineering properties of soil-based controlled low-strength materials as slag partially substitutes to portland cement, *Construction and Building Materials*, 48, 822–829.

Shehata, M.H., Thomas, M.D.A. 2000. The effect of fly ash composition on the expansion of concrete due to alkali–silica reaction, *Cement and Concrete Research*, 30, 1063–1072.

Shehata, M.H., Thomas, M.D.A. 2002. Use of ternary blends containing silica fume and fly ash to suppress expansion due to alkali–silica reaction in concrete, *Cement and Concrete Research*, 32(3), 341–349.

Shi, C., Day, R.L. 2000. Pozzolanic reaction in the presence of chemical activators: Part II – reaction products and mechanisms, *Cement and Concrete Research*, 30(4), 607–613.

Shi, C., Day, R.L. 2001. Comparison of different methods for enhancing reactivity of pozzolans, *Cement and Concrete Research*, 31, 813–818.

Shi, C., Zheng, K. 2007. A review on the use of waste glasses in the production of cement and concrete, resources, *Conservation and Recycling*, 52(2), 234–247.

Shi, Y., Matsui, I., Feng, N. 2002. Effects of compound mineral powders on workability and rheological property of HPC, *Cement and Concrete Research*, 32, 71–78.

Siddique, R. 2010. Use of municipal solid waste ash in concrete, *Resources, Conservation and Recycling*, 55, 83–91.

Siddique, R. 2011. Utilization of silica fume in concrete: Review of hardened properties, *Resources Conservation and Recycling*, 55, 923–932.

Siddique, R., Noumowe, A. 2008. Utilization of spent foundry sand in controlled low-strength materials and concrete, *Resources, Conservation and Recycling*, 53, 27–35.

Singh, M., Kapur, P.C., Pradip, A. 2008. Preparation of alinite based cement from incinerator ash, *Waste Management*, 28, 1310–1316.

Smith, I.A. 1967. The design of fly-ash concretes, *Proceedings of Institution of Civil Engineers*, 36, 769–790.

Smolczyk, H.G. 1980. Slag structure and identification of slag, principal report, *7th International Conference on Chemistry of Cement*, Vol. 1, pp. III/1–17, Paris.

Snellings, R., Mertens, G., Elsen, J. 2012. Supplementary cementitious materials, *Reviews in Mineralogy & Geochemistry*, 74, 211–278.

Song, H-W., Saraswathy, V. 2006. Studies on the corrosion resistance of reinforced steel in concrete with ground granulated blast-furnace slag – an overview, *Journal of Hazardous Materials B1*, 38, 226–233.

Soroka, I., Stern, N. 1976. Calcareous fillers and the compressive strength of portland cement, *Cement and Concrete Research*, 6(3), 367–376.

Sotiriadis, K., Nikolopoulou, E., Tsivilis, S. 2012. Sulfate resistance of limestone cement concrete exposed to combined chloride and sulfate environment at low temperature, *Cement and Concrete Composites*, 34, 903–910.

Sotiriadis, K., Nikolopoulou, E., Tsivilis, S., Pavlou, A., Chaniotakis, E., Swamy, R.N. 2013. The effect of chlorides on the thaumasite form of sulfate attack of limestone cement concrete containing mineral admixtures at low temperatures, *Construction and Building Materials*, 43, 156–164.

Sprung, S., Siebel, E. 1991. Assessment of suitability of limestone for producing portland limestone cement, *Zement-Kalk-Gips*, 44(1), 1–11.

Stanton, T.E. 1940. Expansion of concrete through reaction between cement and aggregate, *Proceedings of American Society of Civil Engineering*, 66(10), 1781–1811.

Stark, D. 1978. Alkali–silica reactivity in the rocky mountain region, *Proceedings of 4th International Conference on the Effect of Alkalies in Cement and Concrete*, Purdue University, 235–243.

Stark, D. 1991. The moisture condition of field concrete exhibiting ASR Reactivity, 973–987, *CANMET/ACI Second International Conference on Durability of Concrete*, SP-126, American Concrete Institute, Detroit.

Stark, D., Morgan, B., Okamoto, P. 1993. Eliminating or minimizing alkali–silica reactivity. *Strategic Highway Research Program*, National Research Council, Washington DC, 266p.

Sullentrop, M.G., Baldwin Jr., J.W. 1983. High-lime fly ash as a cementing agent, *Proceedings of CANMET/ACI 1st International Conference on the Use of Fly Ash, Silica Fume*, Slag, and Other Mineral By-Products in Concrete, Vol. II, pp. 321–331, Montebello.

Sun, G.-K., Young, J.F. 1993. Quantitative determination of residual silica fume in DSP cement pastes by 29-Si NMR, *Cement and Concrete Research*, 23(2), 480–483.

Taha, R.A., Alnuaimi, A.S., Al-Jabri, K.S., Al-Harty, A.S. 2007. Evaluation of controlled low-strength materials containing industrial by-products, *Building and Environment*, 42, 3366–3372.

Takemoto, K., Uchikawa, H. 1980. Hydration of pozzolanic cements, *Proceedings of 7th International Congress on Chemistry of Cement*, Vol. I, Sub-theme IV-2, pp. 1–21, Paris.

Tang, M.S., Liu, Z., Han, S.F. 1986. Mechanism of alkali–carbonate reaction. In *Proceedings of the 7th International Conference on Concrete Alkali–Aggregate Reactions*, Vol. 1, 275–279, Ottawa.

Tarun, R., Singh, S.S. 1997. Influence of fly ash on setting and hardening characteristics of concrete systems, *ACI Materials Journal*, 94(5), 355–360.

Taylor, H.F.W., Famy, C., Scrivener, K.L. 2001. Delayed ettringite formation, *Cement and Concrete Research*, 31, 683–693.

Taylor, R., Richardson, I.G., Brydson, R.M.D. 2010. Composition and microstructure of 20-year-old portland cement-ground granulated blast furnace slag blends containing 0 to 100% slag, *Cement and Concrete Research*, 40(7), 971–983.

Tazawa, E., Miyazawa, S. 1995. Influence of cement and admixture on autogenous shrinkage of cement paste, *Cement and Concrete Research*, 25(2), 281–287.

TÇMB (Turkish Cement Manufacturers' Association) R&D Institute. 1997. Chemical Analysis Report.

TÇMB (Turkish Cement Manufacturers' Association) R&D Institute. 2010a. Chemical Analysis Report.

TÇMB (Turkish Cement Manufacturers' Association) R&D Institute. 2010b. Chemical Analysis Report.

TÇMB (Turkish Manufacturers' Association) R&D Institute. 2014. Chemical Analysis Report.

Temiz, H., Kantarcı, F. 2014. Investigation of durability of CEM II B-M mortars and concrete with limestone powder, calcite powder, and fly ash, *Construction and Building Materials*, 68, 517–524.

Termkhajornkit, P., Nawa, T., Nakai, M., Saito, T. 2005. Effect of fly ash on autogeneous shrinkage, *Cement and Concrete Research*, 35(3), 473–482.

Theodoridou, M., Ioannou, I., Philokyprou, M. 2013. New evidence of early use of artificial pozzolanic material in mortars, *Journal of Archaeological Science*, 40, 3263–3269.

Thomas, M. 1996. Chloride thresholds in marine concrete, *Cement and Concrete Research*, 26(4), 513–519.

Thomas, M. 2013. *Supplementary Cementing Materials in Concrete*. CRC Press, Boca Raton, FL.

Thomas, M., Fournier, B., Folliard, K., Ideker, J., Shehata, M. 2006. Test methods for evaluating preventive measures for controlling expansion due to alkali–silica reaction in concrete, *Cement and Concrete Research*, 36, 1842–1856.

Thomas, M.D.A., Shehate, M.H., Shashiprakash, S.G., Hopkins, D.S., Cail, K. 1999. Use of ternary cementitious systems containing silica fume and fly ash in concrete, *Cement and Concrete Research*, 29, 1207–1214.

Tian, B., Cohen, M.D. 2000. Does gypsum formation during sulfate attack on concrete lead to expansion? *Cement and Concrete Research*, 30, 117–123.

Tikalsky, P., Gaffney, M., Regan, R. 2000. Properties of controlled low-strength material containing foundry sand, *ACI Materials Journal*, 97(6), 698–702.

Tikalsky, P.J., Carasquillo, R.L. 1993. Fly ash evaluation and selection for use in sulfate resistant concrete, *ACI Materials Journal*, 90(6), 545–551.

Tishmack, J.K., Olek, J., Diamond, S., Sahu, S. 2001. Characterization of pore solutions expressed from high-calcium fly ash-water pastes, *Fuel*, 80(6), 815–819.

Tokyay, M. 1987. Effect of a high-calcium fly ash and a low-calcium fly ash on the properties of portland cement fly ash pastes and mortar, PhD thesis, Middle East Technical University, Ankara, Turkey, 170pp.

Tokyay, M. 1999. Strength prediction of fly ash concretes by accelerated testing, *Cement and Concrete Research*, 29, 1737–1741.

Tokyay, M., Delibaş, T., Aslan, Ö. 2010. Effects of mineral admixture type, *Grinding Process, and Cement Fineness on the Physical and Mechanical Properties of GGBFS-, Natural Pozzolan-, and Limestone-incorporated Cements*, Working Paper AR-GE 2010/01-B, Turkish Cement Manufacturers' Association (TÇMB), 45p. (in Turkish).

Tokyay, M., Dilek, F.T. 2003. Sulfate resistance of cementitious systems with mineral additives. Final Report of Project No: 1991010, 243 pp. Granted by TÜBİTAK, The Scientific and Technical Research Council of Turkey, Construction and Environmental Technologies Research Grant Committee.

Tokyay, M., Erdoğdu, K. 1998, *Characterization of Turkish Fly Ashes*, TÇMB/AR-GE/Y 98.3, Turkish Cement Manufacturers' Association, 70pp. (in Turkish).

Tokyay, M., Hubbard, F.H. 1992. Mineralogical investigations of high-lime fly ashes, *4th CANMET/ACI International Conference on. FlyAsh, Silica Fume, Slag and Natural Pozzolans in Concrete* (Ed. V.M. Malhotra), Vol. 1, pp. 65–78, Istanbul.

Tonak, T. 1995. Usability of diatomaceous earth wastes in cement industry (in Turkish), *Cement Bulletin (TÇMB)*, 28, 291–297.

Tong, L., Tang, M. 1999. Expansion mechanism of alkali–dolomite and alkali–magnesite reaction, *Cement and Concrete Composites*, 21, 361–373.

Tong, Y., Du, H., Fei, L. 1991. Comparison between the hydration processes of tricalcium silicate and beta-dicalcium silicate, *Cement and Concrete Research*, 21(4), 509–514.

Torgal, F.P., Miraldo, S., Labrincha, J.A., De Brito, J. 2012. An overview on concrete carbonation in the context of eco-efficient construction: Evaluation, use of SCMs and/or RAC, *Construction and Building Materials*, 36, 141–150.

Tosun, K. 2006. Effect of SO_3 Content and fineness on the rate of delayed ettringite formation in heat cured portland cement mortars, *Cement and Concrete Composites*, 28, 761–772.

Tosun, K. 2007. Effects of different cement types on delayed ettringite formation, PhD thesis, Graduate School of Natural and Applied Sciences, Dokuz Eylül University, İzmir.

Tosun, K., Felekoğlu, B., Baradan, B., Altun, I.A. 2009. Portland limestone cement Part I- preparation of cements, *Teknik Dergi*, 20(3), 4717–4736.

TS 802. 2005. Design Criteria for Concrete Mixtures. Turkish Standards Institute, Ankara.

TS EN 12390-8. 2010. Testing Hardened Concrete – Part 8: Depth of Penetration of Water Under Pressure. Turkish Standards Institute, Ankara.

Tsivilis, S., Chaniotakis, E., Badogiannis, E., Pahoulas, G., Ilias, A. 1999. A study on the parameters affecting the properties of portland limestone cements, *Cement and Concrete Composites*, 21, 107–116.

Tsivilis, S., Tsantilas, J., Kakali, G., Chaniotakis, E., Sakellariou, A. 2003. The permeability of portland limestone cement concrete, *Cement and Concrete Research*, 33, 1465–1471.

Tuan, N.V., Ye, G., van Breugel, K., Fraaij, A.L.A., Dai, B.D. 2011. The study of using rice husk ash to produce ultra high-performance Concrete, *Construction and Building Materials*, 25, 2030–2035.

Tumidaski, P.J., Thomson, M.L. 1994. Influence of cadmium on hydration of C3A, *Cement and Concrete Research*, 24(7), 1359–1372.

Turanlı, L., Bektaş, F., Monteiro, P.J.M. 2003. Use of ground clay brick as a pozzolanic material to reduce the alkali–silica reaction, *Cement and Concrete Research*, 33(10), 1359–1362.

Turhan, Ş., Parmaksız, A., Köse, A., Yüksel, C., Arıkan, İ.H., and Yücel, B. 2010. Radiological characteristics of pulverized fly ashes produced in Turkish coal-burning thermal power plants, *Fuel*, 89, 3892–3900.

Türkcan, M. 1971. Concrete strength and water-cement relations for various types of western Turkish portland cements, MS thesis, Middle East Technical University, Ankara, Turkey, 76pp.

Türker, P., Erdoğan, B., Katnaş, F., Yeğinobalı, A. 2004. Classification and properties of Turkish fly ashes, TÇMB/AR-GE/Y03.03, Turkish Cement Manufacturers' Association, 99pp. (in Turkish).

Tuutti, K. 1982. Corrosion of Steel in Concrete. CBI Research 4.82, Swedish Cement and Concrete Research Institute, Stockholm.

Uchikawa, H., Uchida, S. 1980. Influence of pozzolana on the hydration of C3A, *Proceedings of 7th International Congress on Chemistry of Cement*, Vol. 3, pp. IV/24-IV/29, Paris.

Ucrainczyk, N., Mihelj, N.F., Šipušić, J. 2013. Calcium sulfoaluminate eco-cement from industrial waste, *Chemical and Biochemical Engineering, Quarterly*, 27(1), 83–93.

Ün, H., Baradan, B. 2011. The Effect of curing temperature and relative humidity on the strength development of portland cement mortar, *Scientific Research and Essays*, 6(12), 2504–2511.

Uysal, M., Sümer, M. 2011. Performance of self-compacting concrete containing different mineral admixtures, *Construction and Building Materials*, 25, 4112–4120.

Uzal, B. 2002. Effects of high volume natural pozzolan addition on the properties of pozzolanic cements, MS thesis, Middle East Technical University, Ankara, Turkey, 98pp.

Uzal, B., Turanlı, L., Yücel, H., Göncüoğlu, M.C., Çulfaz, A. 2010. Pozzolanic activity of clinoptilolite: A comparative study with silica fume, fly ash, and a non-zeolitic natural pozzolan, *Cement and Concrete Research*, 40(3), 398–404.

Valenza II, J.J., Scherer, G.W. 2007a. A review of salt scaling: I. phenomenology, *Cement and Concrete Research*, 37, 1007–1021.

Valenza II, J.J., Scherer, G.W. 2007b. A review of salt scaling: II. Mechanisms, *Cement and Concrete Research*, 37, 1022–1034.

Valipour, M., Pargar, F., Shekarchi, M., Khani, S. 2013. Comparing a natural pozzolan, zeolite to metakaolin and silica fume in terms of their effect on the durability characteristics of concrete: A laboratory study, *Construction and Building Materials*, 41, 879–888.

Van den Heede, P., Furniere, J., De Belie, N. 2013. Influence of air-entraining agents on de-icing salt scaling resistance and transport properties of high-volume fly ash concrete, *Cement and Concrete Composites*, 37, 293–303.

Venkatanarayanan, H.K., Rangaraju, P.R. 2013. Decoupling the effects of chemical composition and fineness of fly ash in mitigating alkali–silica reaction, *Cement and Concrete Composites*, 43, 54–68.

Vijayalakshmi, M., Sekar, A.S.S., Ganesh prabhu, G. 2013. Strength and durability properties of concrete made with granite industry waste, *Construction and Building Materials*, 46, 1–7.

Vikan, H., Justnes, H. 2007. Rheology of cementitious paste with silica fume or limestone, *Cement and Concrete Research*, 37, 1512–1517.

von Berg, W., Feuerborn, H.-J. 2005. Present situation and perspectives of CCP management in Europe, *2005 World of Coal Ash (WOCA) Conference*, Lexington, Kentucky, USA, 11–15.

von Berg, W., Kukko, H. 1991. Fresh mortar and concrete with fly ash, in *Fly Ash in Concrete – Properties and Performance, Rept. Tech. Comm. 67-FAB RILEM* (Ed. K. Wesche), E.&F.N. Spon, London.

Wainwright, P.J. 1986. Properties of fresh and hardened concrete incorporating slag cements, in *Cement Replacement Materials* (Ed. R.N. Swamy), Surrey University Press, Guildford, Surrey, UK, pp. 100–133.

Weeks, C., Hand, R.J., Sharp, J.H. 2008. Retardation of cement hydration caused by heavy metals present in ISF slag used as aggregate, *Cement and Concrete Composites*, 30, 970–978.

Wei, S., Handong, Y., Binggen, Z. 2003. Analysis of mechanism on water-reducing effect of fine ground slag, high-calcium fly ash, and low-calcium fly ash, *Cement and Concrete Research*, 33, 1119–1125.

Whitfield, C.J. 2003. The production, disposal, and beneficial use of coal combustion products in Florida, MS thesis, University of Florida, USA, 176pp.

Wigum, B.J. 1995. Alkali–aggregate reactions in concrete, PhD thesis, University of Trondheim, The Norwegian Institute of Technology, Department of Geology and Mineral Resources Engineering, Trondheim.

Wild, S., Sabir, B.B., Khatib, J.M. 1995. Factors influencing strength development of concrete containing silica fume, *Cement and Concrete Research*, 25(7), pp. 1567–1580.

Wong, H.S., Abdul Razak, H. 2005. Efficiency of calcined kaolin and silica fume as cement replacement material for strength performance, *Cement and Concrete Research*, 35, 696–702.

Wong, Y.L., Lam, L., Poon, C.S., Zhou, F.P. 1999. Properties of fly ash-modified cement mortar-aggregate interface, *Cement and Concrete Research*, 29, 1905–1913.

Wright, J.R., Shafaatian, S., Rajabipour, F. 2014. Reliability of chemical index model in determining fly ash effectiveness against alkali–silica reaction induced by highly reactive glass aggregates, *Construction and Building Materials*, 64, 166–171.

Wu, K., Shi, H., De Schutter, G., Guo, X., Ye, G. 2012. Preparation of alinite cement from municipal solid waste incineration fly ash, *Cement and Concrete Composites*, 34, 322–327.

Xu, X., Zhang, Y., Li, S. 2012. Influence of different localities phosphorus slag powder on the performance of portland cement, *Procedia Engineering*, 27, 1339–1346.

Yan, D.Y.S., Tang, I.V., Lo, I.M.C. 2014. Development of controlled low-strength material derived from beneficial reuse of bottom ash and sediment for green construction, *Construction and Building Materials*, 64, 201–207.

Yazıcı, H., Yardımcı, M.Y., Aydın, S., Karabulut, A.Ş. 2009. Mechanical properties of reactive powder concrete containing mineral admixtures under different curing regimes, *Construction and Building Materials*, 23, 1223–1231.

Yazıcı, H., Yiğiter, H., Aydın, S., Baradan, B. 2006. Autoclaved SIFCON with high volume class C fly ash binder phase, *Cement and Concrete Research*, 36, 481–486.

Yeğinobalı, A. 2009., Silica fume. Its use in cement and concrete, TÇMB/AR-GE/Y01.01, Turkish Cement Manufacturers' Association, 62pp. (in Turkish).

Yiğiter, H., Aydın, S., Yazıcı, H., Yardımcı, M.Y. 2012. Mechanical performance of low-cement reactive powder concrete (LCRPC), *Composites: Part B*, 43, 2907–2914.

Yılmaz, A. 1998. Effect of clinker composition on the properties of pozzolanic cements, MS thesis, Middle East Technical University, Ankara, Turkey, 103pp.

Younsi, A., Turcry, Ph., Aït-Mokhtar, A., Staquet, S. 2013. Accelerated carbonation of concrete with high content of mineral additions: Effect of interactions between hydration and drying, *Cement and Concrete Research*, 43, 25–33.

Yunsheng, Z., Wei, S., Sifeng, L., Chujie, J., Jianzhong, L. 2008. Preparation of C200 green reactive powder concrete and its static-dynamic behaviors, *Cement and Concrete Composites*, 30, 831–838.

Zain, M.F.M., Islam, M.N., Jamil, M.M. 2011. Production of rice husk ash for use in concrete as a supplementary cementitious material, *Construction and Building Materials*, 25, 798–805.

Zelić, J. 2005. Properties of concrete pavements prepared with ferrochromium slag as concrete aggregate, *Cement and Concrete Research*, 35, 2340–2349.

Zenati, A., Arroudj, K., Lanez, M., Oudjit, M.N. 2009. Influence of cementitious additions on rheological and mechanical properties of reactive powder concretes, *Physics Procedia*, 2, 1255–161.

Zhang, C., Wang, A., Tang, M., Wu, B., Zhang, N. 1999. Influence of aggregate size and aggregate size grading on ASR expansion, *Cement and Concrete Research*, 29, 1393–1396.

Zhang, M., Chen, J., Lv, Y., Wang, D., Ye, J. 2013. Study on the expansion of concrete under attack of sulfate and sulfate-chloride ions, *Construction and Building Materials*, 39, 26–32.

Zhang, M.H., Malhotra, V.M. 1996. High-performance concrete incorporating rice husk ash as a supplementary cementing material, *ACI Materials Journal*, 93(6), 629–636.

Zhang, T., Yu, Q., Wei, J., Zhang, P. 2012. Efficient utilization of cementitious materials to produce sustainable blended cement, *Cement and Concrete Composites*, 34, 692–699.

Zhang, Z., Provis, J.L., Reid, A., Wang, H. 2014. Fly ash-based geopolymers: The relationship between composition, pore structure and efflorescence, *Cement and Concrete Research*, 64, 30–41.

Zhang, Z., Qian, S., Ma, H. 2014. Investigating mechanical properties and self-healing behavior of micro-cracked ECC with different volume of fly ash, *Construction and Building Materials*, 52, 17–23.

Zhu, Y., Yang, Y., Yao, Y. 2012. Use of slag to improve mechanical properties of engineered cementitious composites (ECCs) with high volumes of fly ash, *Construction and Building Materials*, 36, 1076–1081.

Zhu, Y. Zhang, Z., Yang, Y., Yao, Y. 2014. Measurement and correlation of ductility and compressive strength for engineered cementitious composites (ECC) produced by binary and ternary systems of binder materials: Fly ash, slag, silica fume and cement, *Construction and Building Materials*, 68, 192–198.

Zollo, R.F. 1997. Fiber-reinforced concrete: An overview after 30 years of development, *Cement and Concrete Composites*, 19, 107–122.

Index